Blood Gases and Critical Care Testing
Physiology, Clinical Interpretations, and Laboratory Applications

Blood Gases and Critical Care Testing

Physiology, Clinical Interpretations, and Laboratory Applications

Third Edition

John G. Toffaletti, PhD
Department of Pathology
Clinical Laboratories
Duke Medical Center
Chief of Clinical Chemistry
VA Medical Center
Durham, NC, United States

Craig R. Rackley, MD
Department of Medicine
Division of Pulmonary, Allergy, and Critical Care
Duke University Medical Center
Durham, NC, United States

Published in cooperation with AACC

ACADEMIC PRESS
An imprint of Elsevier

ELSEVIER

Academic Press is an imprint of Elsevier
125 London Wall, London EC2Y 5AS, United Kingdom
525 B Street, Suite 1650, San Diego, CA 92101, United States
50 Hampshire Street, 5th Floor, Cambridge, MA 02139, United States
The Boulevard, Langford Lane, Kidlington, Oxford OX5 1GB, United Kingdom

Notices
Knowledge and best practice in this field are constantly changing. As new research and
experience broaden our understanding, changes in research methods, professional
practices, or medical treatment may become necessary.

Practitioners and researchers must always rely on their own experience and knowledge in
evaluating and using any information, methods, compounds, or experiments described
herein. In using such information or methods they should be mindful of their own safety
and the safety of others, including parties for whom they have a professional responsibility.

To the fullest extent of the law, neither the Publisher nor the authors, contributors, or
editors, assume any liability for any injury and/or damage to persons or property as a matter
of products liability, negligence or otherwise, or from any use or operation of any methods,
products, instructions, or ideas contained in the material herein.

About AACC
Dedicated to achieving better health through laboratory medicine, AACC brings together
more than 50,000 clinical laboratory professionals, physicians, research scientists, and
business leaders from around the world focused on clinical chemistry, molecular diagnostics,
mass spectrometry, translational medicine, lab management, and other areas of progressing
laboratory science. Since 1948, AACC has worked to advance the common interests of the field,
providing programs that advance scientific collaboration, knowledge, expertise, and
innovation. For more information, visit www.aacc.org.

Published in cooperation with AACC

Library of Congress Cataloging-in-Publication Data
A catalog record for this book is available from the Library of Congress

British Library Cataloguing-in-Publication Data
A catalogue record for this book is available from the British Library

ISBN: 978-0-323-89971-0

For information on all Academic Press publications visit our website at
https://www.elsevier.com/books-and-journals

Publisher: Stacy Masucci
Acquisitions Editor: Ana Claudia A. Garcia
Editorial Project Manager: Mona Zahir
Production Project Manager: Kiruthika Govindaraju
Cover Designer: Miles Hitchen

Typeset by TNQ Technologies

Working together
to grow libraries in
developing countries

www.elsevier.com • www.bookaid.org

Contents

About the authors

John Toffaletti

John Toffaletti received a BS from the University of Florida in Gainesville and followed this with training in clinical chemistry at the University of North Carolina at Chapel Hill, where he earned a PhD in Biochemistry under Drs. John Savory and Hillel Gitelman, and then completed a Postdoctoral Fellowship in Clinical Chemistry at Hartford Hospital under Dr. George Bowers, Jr.

Since completing these programs, he has worked in the Clinical Laboratories at Duke University Medical Center since 1979, where he is now Professor of Pathology, Director of the Blood Gas Laboratory, the Clinical Pediatric Laboratory, and several Outpatient Laboratories. He is also the Chief of Clinical Chemistry at the Durham VA Medical Center.

He has written or presented numerous workshops, books, study guides, chapters, and seminars on the interpretation of blood gas, cooximetry, ionized calcium, magnesium, lactate (sepsis), kidney function tests (creatinine, cystatin C, GFR), and viscoelastic testing (ROTEM and TEG). His research interests include sample collection, preanalytical errors, analysis, and clinical use of these tests.

He has served on numerous committees in his area and served on several boards of editors of journals, including the Board of Editors of Clinica Chimica Acta since 1999. In addition to his numerous scientific presentations, he has chaired several scientific program committees at national and international meetings.

Among his outside interests are bodyboarding, furniture making, and cycling, in which he has a patented design for a road handlebar. Also, after mostly not playing for nearly 40 years, he restarted playing the piano about 10 years ago and now regularly plays the Steinway at the Duke Cancer Center.

Craig Rackley

Craig Rackley received a BS in biochemistry and molecular biology from Oklahoma State University and then went on to medical school at Georgetown University School of Medicine in Washington, DC. He completed his residency training in internal medicine at the University of California, San Francisco and then his fellowship in pulmonary and critical care medicine at Duke University Medical Center.

Upon completion of his training, he joined the faculty at Duke University in 2013. He is currently an Associate Professor of Medicine and is the medical director of Duke's adult extracorporeal membrane oxygenation program. He teaches in the school of medicine and practices both outpatient pulmonary medicine and inpatient critical care medicine.

He has written articles and chapters on gas exchange, lung injury, mechanical ventilation, and extracorporeal life support. Furthermore, he frequently serves as a faculty at national and international training courses on mechanical ventilation and extracorporeal life support.

Outside of medicine, his interests include spending time with his family, reading, and vegetable gardening.

Preface to third edition

This is now the third edition of my book that was titled *Blood Gases and Electrolytes*. The dramatic changes in the world of critical care testing were the impetus for a new edition with a new title. Indeed, the first edition of the book seems from a different universe. As I recall, the main topic was how to measure phlogiston on a Van Slyke apparatus and how to treat hypophlogistonemia. The second edition was better, but critical care testing has continued to change in the past 12 years since, so this third edition is a thoroughly updated version deserving a new title. In addition to updated chapters on clinical interpretations of blood gases and electrolytes (Ca, Mg, Na, K, and PO4), it includes entirely new chapters on blood gas testing on venous, capillary and umbilical cord blood, the evolving role of blood lactate testing in critical care, proper sample collection and handling to avoid preanalytical variation, and chapters on point-of-care testing and quality control for these tests.

A most significant addition has been having Dr. Craig Rackley as a coauthor. Dr. Rackley is a critical care pulmonologist at Duke Medical Center who is able to add the physician's perspective to critical care testing. He is an excellent speaker and writer, and we have collaborated with several presentations at national meetings.

Chapter 1

Introduction to blood-gas tests and blood-gas physiology

Introduction and history of blood gases

The term "blood gas" refers to the parameters pH, pCO_2, and pO_2 measured in blood. Note that the little "p" in pH stands for negative log, while the italicized p in pCO_2 and pO_2 stands for the partial pressure of each of these gases. In addition to pH, pCO_2, and pO_2, modern "blood gas analyzers" may also measure the hemoglobin fractions, electrolytes, and metabolites such as sodium, potassium, ionized calcium, chloride, bicarbonate, glucose, and lactate. Some analyzers also include measurements of creatinine, urea, and ionized magnesium.

The history of blood gases and oximetry has perhaps the oldest, best documented, and, to some of us, the most interesting history of developments in laboratory tests. The history includes many notable figures, including Joseph Priestley, who became fascinated observing the large volume of gas produced in making beer and then went on to isolate 10 different gases, including oxygen in the late 1700s. Around that time, the eccentric and exceedingly wealthy (by an unexpected inheritance) Henry Cavendish, once described as *"the richest of all learned men, and probably also the most learned of all the rich,"* discovered hydrogen, characterized carbon dioxide, and was the first person to accurately analyze atmospheric air. The early history of blood gases even includes Benjamin Franklin, a colleague of many scientists including Priestley and a founding father of the United States of America. To paraphrase Alan Grogono: *"In addition to publishing newspapers, drafting constitutions, serving as postmaster general, flying kites in thunderstorms, discovering the Gulf Stream, and maintaining friendships with French ladies, Benjamin Franklin found time to make an unfortunate guess about calling "vitreous" charges "positive."* This decision later led to assigning a "negative" charge to electrons and a "positive" charge to hydrogen ions *(1)*.

Several distinguished scientists have contributed to the definition of an acid. In the late 1800s, Arrhenius defined acids as hydrogen salts. In the early 1900s, Lawrence Henderson and Karl Hasselbalch sequentially characterized the buffering relationship between an acid and base thereby creating the eponymous Henderson–Hasselbalch equation, but who never actually knew

Blood Gases and Critical Care Testing. https://doi.org/10.1016/B978-0-323-89971-0.00002-1

1

each other. Brønsted and Lowry simultaneously, but separately, defined acids as substances that could donate a hydrogen ion, and Gilbert Lewis later described an acid as any compound that could accept a pair of electrons to form a covalent bond.

Donald Van Slyke embraced the idea that acid–base status was partly determined by electrolytes, an idea that was expanded by Peter Stewart into the very complex Strong-Ion-Difference explanation of acid–base balance *(2)*. Importantly, Van Slyke is credited with expanding chemical analyses into the hospital and is considered a founder of "clinical chemistry." One of his most notable discoveries was the gasometric method, which measured released O_2 gas and consequentially the oxygen saturation in the blood. Before and during World War II, Kurt Kramer, J.R. Squires, and Glen Millikan made significant advancements in oximetry, which led to its integrated use with oxygen-delivery systems enabling safer high-altitude military flights. These developments eventually led Takuo Aoyagi to the discovery of pulse oximetry in the 1970s, which allows for the separation of the arteries absorption of hemoglobin from the absorption of the tissue using the pulsatile nature of the arterial absorption signal.

The prototype electrode for measuring the partial pressure of oxygen (pO_2) was developed in 1954 by Leland Clark, using polyethylene film and other materials that cost less than a dollar. Also in 1954, Richard Stowe covered a pH electrode with a rubber finger covering to develop a prototype of today's partial pressure of carbon dioxide (pCO_2) electrodes. These stories and many others that led to the development of the blood-gas analyzer were documented in a book by Astrup and Severinghaus *(3)*.

The methodology for measuring clinical blood gases has evolved dramatically from mostly large laboratory-dedicated analyzers to hybrid analyzers adaptable to both laboratory and near-patient settings. There has also been a huge growth in the use of portable hand-held analyzers that are suited for smaller laboratories and near-patient use in hospitals, clinics, or remote locations. Blood gas testing is widely used as a tool for diagnosing disorders and evaluating the efficacy of therapeutic interventions.

Explanations of blood gas, acid–base, and cooximetry terms

pH. pH is an index of the acidity or alkalinity of the blood. Normal arterial pH is 7.35–7.45. A pH <7.35 indicates an acid state, and a pH >7.45 indicates an alkaline state. Acidemia refers to the condition of the blood being too acidic, and acidosis refers to the metabolic or respiratory process within the patient that causes acidemia. The adjective for the process is acidotic. Similar terms are used for the alkaline state: alkalemia, alkalosis, and alkalotic. Because all enzymes and physiological processes may be affected by pH, pH is normally regulated within a very tight physiologic range, especially within an individual, but also for reference intervals (see Table 1.1).

TABLE 1.1 Reference intervals for venous and arterial blood.

Test	Age category	Ref (36) Arterial	Duke medical center Venous	Duke medical center Arterial	Ref (32) Venous	Ref (32) Arterial	Ref (31,33) Arterial
pH	1–2 h	7.35–7.45					
pH	20–76 years						7.38–7.46
pH	60–90 years		7.32–7.42	7.35–7.45	7.31–7.41	7.35–7.45	
pCO_2 mmHg	1–2 h	32–45					
pCO_2 mmHg	20–76 years		39–55	35–45	41–51	35–45	32–45 (F) 35–48 (M)
pO_2 mmHg	1–2 h	65–96					
pO_2 mmHg	20–39 years						83–115
pO_2 mmHg	40–76 years						70–110
pO_2 mmHg	2 days–60 years		30–55	75–108	30–40	80–100	
sO_2 (%)					~75	>95	
Base excess			–3.0 to +3.0				–3.0 to +2.8
HCO_3 mmol/L Total CO_2	2 years–adult		21–30	21–29	23–29	22–28	21–27
Anion gap mmol/L			5–11				

***p*CO₂.** *p*CO$_2$ is a measure of the tension or pressure of carbon dioxide dissolved in the blood. The *p*CO$_2$ of blood represents the balance between cellular production and diffusion of CO$_2$ into the blood and ventilatory removal of CO$_2$ from blood. A normal, steady *p*CO$_2$ indicates that the lungs are removing CO$_2$ from blood at about the same rate as CO$_2$ produced in the tissues is diffusing into the blood. A change in *p*CO$_2$ indicates an alteration in this balance. CO$_2$ is an acidic gas that is largely controlled by our rate and depth of breathing or ventilation, which changes to match the rate of metabolic production of CO$_2$. *p*CO$_2$ is classified as the respiratory or ventilatory component of acid—base balance.

***p*O₂.** *p*O$_2$ is a measure of the tension or pressure of oxygen dissolved in the blood. The *p*O$_2$ of arterial blood is primarily related to the ability of oxygen to enter the lungs and diffuse across the alveoli into the blood. As shown in Fig. 1.1, there is a continual gradient of *p*O$_2$ from atmospheric air (150 mmHg) to the alveoli ($\sim$110 mmHg), to arterial blood ($\sim$100 mmHg), capillaries ($\sim$60 mmHg) and venous blood ($\sim$40 mmHg), and finally to the mitochondria in cells with the lowest *p*O$_2$ of $\sim$8—12 mmHg. These gradients drive the movement, binding, and release of oxygen among these systems.

Common causes of a decreased arterial *p*O$_2$ are listed below, with further details presented in Chapter 4:

- Hypoventilation: Alveolar ventilation is low relative to O$_2$ uptake and CO$_2$ production, which leads to decreased alveolar *p*O$_2$ and increased alveolar *p*CO$_2$. Example: severe obstructive lung disease.
- Low oxygen environment: Partial pressure of oxygen in inspired air is less than 160 mmHg. This is most commonly seen at high altitudes.
- Ventilation/perfusion (V/Q) mismatch: Areas within the lung are receiving inspired air, but the perfusion to that portion of the lung is limited. Example: Pulmonary embolism, where a clot lodges in a pulmonary artery to limit blood flow to an otherwise functioning lung unit.

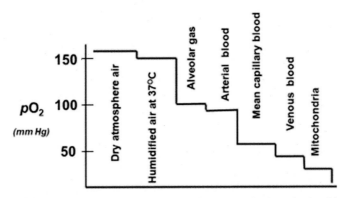

FIGURE 1.1 Gradients of *p*O$_2$ from atmospheric air to blood to mitochondria.

- Shunt: A portion of the blood travels from the venous system to the arterial system without contact with a functioning alveolar unit. Example: Lung disease where blood flows through portions of the lung that are unventilated due to complete airway obstruction, atelectasis, or filling with fluid or cells.
- Diffusion impairment: Oxygen is unable to efficiently transfer across the blood-gas barrier of the alveoli. Examples: Thickening of the blood-gas barrier due to fibrosis, edema, or inflammatory cell infiltration.

Bicarbonate. Although the bicarbonate ion (HCO_3^-) can now be measured directly, some blood-gas analyzers calculate [HCO_3^-] with the Henderson−Hasselbalch equation from measurements of the pH and pCO_2. Bicarbonate is an indicator of the buffering capacity of blood and is classified as the metabolic component of acid−base balance.

Base Excess. Base excess (BE) is a calculated term that describes the amount of bicarbonate relative to pCO_2. Standard BE reflects only the metabolic component of acid−base disturbances. It is based on the titratable fluid volume throughout the body (both extravascular and vascular (blood)) and also includes the contribution of Hb for acid−base disturbances.

The BE concept has a long history with some spirited controversy that is reviewed by Johan Kofstad, a professional colleague and most gracious friend who I met in Oslo on my first trip to Europe in 1983 *(5)*. Base excess was conceived by Astrup in the 1950s and refined with equations and nomograms by Siggaard-Andersen in 1960. In 1977, attempting to resolve controversies about BE between the Americans and Denmark, Severinghaus proposed a modified nomogram. However, the different beliefs related to whether BE should be calculated in blood or extracellular fluid remain unreconciled to this day.

Equations for calculating extracellular base excess (BE_{ECF}) from pH and either HCO_3 (mmol/L) or pCO_2 (mmHg) appear to be different but are eerily similar. Here are two equations used for calculating BE *(4,6)*:

$$BE_{ECF} = 16.2 \times (pH - 7.40) - 24.8 + \left[HCO_3^- \right] \qquad (1.1)$$

$$BE_{ECF} = 0.028 \times pCO_2 \times 10^{(pH-6.1)} + 13.8 \times pH - 124.6 \qquad (1.2)$$

The "normal" reference interval for BE is −3 to +3 mmol/L. Comparison of the calculated BE to the reference range for BE may help determine whether an acid/base disturbance is a respiratory, metabolic, or mixed metabolic/respiratory problem. A base excess value exceeding +3 indicates metabolic alkalosis such that the patient requires increased amounts of acid to return the blood pH to neutral if pCO_2 is normal. A base excess below −3 indicates metabolic acidosis and excess acid needs to be removed from the blood to return the pH back to neutral if pCO_2 is normal. My personal opinion is that the BE calculation adds little to the simple interpretation of the difference of the measured bicarbonate −24, especially for pH from 7.3 to 7.5, as noted below and in Table 1.2.

TABLE 1.2 Comparisons of base excess to bicarbonate difference.

pH	HCO_3	pCO_2	BE	$HCO_3 - 24$
7.2	11	30	−14	−13
7.2	15	40	−11	−9
7.2	19	50	−8	−5
7.2	23	60	−4	−1
7.3	14	30	−11	−10
7.3	19	40	−6	−5
7.3	24	50	−2	0
7.4	18	30	−6	−6
7.4	24	40	−0.4	0
7.4	30	50	5	6
7.5	23	30	−0.3	−1
7.5	30	40	7	6
7.5	38	50	14	14
7.6	19	20	−2	−5
7.6	29	30	7	5
7.6	38	40	15	14

When the value of the BE is a negative number, it is frequently referred to as a base deficit (BD). The BD is often used to guide resuscitation in patients suffering from shock where hypoperfusion leads to inadequate delivery of oxygen to the tissue resulting in metabolic acidosis *(7)*. As the patient is successfully resuscitated and oxygen delivery restored, the BD will begin to normalize. In patients who have undergone acute physical trauma, the BD has a significant prognostic value *(8)*.

Table 1.2 shows that the relationship between BE and the simple difference of (measured bicarbonate—24 mmol/L) is usually 2 mmol/L or less, especially for pH from 7.3 to 7.5 as noted earlier. We leave it to each reader to determine the clinical importance of calculating BE versus the use of the bicarbonate concentration.

Anion gap

The anion gap (AG) is a calculated term for the difference between the commonly measured cations (Na^+ and sometimes K^+) and the commonly measured anions (Cl^- and HCO_3^-). Therefore, it represents the unmeasured anions such as negatively charged proteins (particularly albumin) and lactate, phosphates, sulfates, urates, and ketones produced by the body. Exogenous toxins and drugs, including methanol, salicylate, and ethylene glycol (and its metabolites), also contribute to the anion gap when present. The AG is calculated as follows:

$$AG = [Na^+] - [Cl^-] - [HCO_3^-] \quad \text{(Reference Range: } 8 - 16 \text{ mmol} / L)$$
$$(1.3)$$

If K^+ is included in the calculation, the formula is as follows:

$$AG = [Na^+] + [K^+] - [Cl^-] - [HCO_3^-] \quad \text{(Reference Range: } 12 - 20 \text{ mmol} / L)$$
$$(1.4)$$

Correcting the AG for abnormal albumin concentrations is important as it is the greatest contributor to the AG in health. Generally, for each g/dL decrease in albumin, the AG is reduced by about 2.5 mmol/L (or 0.25 mmol/L for each g/L decrease in albumin) or is similarly increased for a rise in serum albumin (9,10). The AG is useful in diagnosing a metabolic acidosis and differentiating among the causes, as shown in Table 1.3.

While often very useful, AG is calculated from three or four measurements and is subject to a variation of up to ±4 mmol/L. Consequentially, it may detect elevated lactate in only about half of the cases (9). Even in patients with an AG in the range of 20−29 mmol/L, only about two-thirds will have a metabolic acidosis, while all patients with an AG higher than this will have a metabolic acidosis (11,12).

Delta gap and delta ratio

The "delta gap" is the difference between the increase in AG and the decrease in HCO_3, while the "delta ratio" is the ratio of the increase in AG divided by

TABLE 1.3 Changes of anion gap in various acid−base disorders.

Disorder	Decreased	Gained	Effect on AG
Diarrhea	HCO_3^-	Cl	Little
Renal tubular acidosis	HCO_3^-	Cl	Little
Lactate acidosis	HCO_3^-	Lactate	Increased
Ketoacidosis	HCO_3^-	Ketoacids	Increased
Mixed disorder: ketoacidosis with metabolic alkalosis	HCO_3^-	Ketoacids and HCO_3^-	Increased (with little change in total CO_2)

the decrease in HCO_3. They can help determine if the high anion gap metabolic acidosis is solely explained by the decrease in HCO_3, or if a possible mixed acid−base disorder is present. These calculations and their application will be discussed in Chapter 2.

Strong ion difference

The strong ion difference (SID) is a concept developed by Peter A Stewart in 1981 aimed at explaining pH changes and at assessing clinical acid−base disturbances *(2,13,14)*. It is based on differences between the strongly dissociated cations Na^+, K^+, Ca^{++}, Mg^{++}, and strongly dissociated anions Cl^- and lactate. Stewart used electroneutrality and the degree of dissociation (strongly or weakly) of electrolytes to explain acid−base physiology. Essentially, the sum of positive charges must be equal to the sum of negative charges. To maintain electroneutrality, the SID concept says that the concentrations of H^+, OH^-, HCO_3^-, and a variety of other weak acids and bases (and therefore the pH) depend on the difference between strongly dissociated cations (like Na^+ and K^+) and anions (such as Cl^- and lactate). A higher SID favors an alkaline environment and a lower SID favors an acidic environment.

The SID concept is complex mathematically, not easily grasped chemically, and is subject to analytical variability because multiple measurements are used in calculating the SID. Even more confusing is that multiple equations and analytes seem to be used in calculating the SID. Some relatively clear equations for SID are *(14,15)*:

$$SID = [Na^+] + [K^+] - [Cl^-] - [lactate] \qquad (1.5)$$

$$SID = [Na^+] + [K^+] + [Ca^{++}] + [Mg^{++}] - [Cl^-] - [lactate] \qquad (1.6)$$

There are also programs that calculate the SID, with one requiring up to 11 results in the calculation, including pH, pCO_2, Na^+, K^+, Cl^-, HCO_3^-, lactate, phosphate, and albumin *(16)*. Another report describes a strong ion calculator that minimally requires $[Na^+]$, $[K^+]$, $[Cl^-]$, pH, and pCO_2 *(17)*. Examples of SID in different solutions are shown in Table 1.4.

Because Stewart considered albumin the most important weak acid *(2)*, Story, et al. *(13)* proposed some relatively simple (but still hard to grasp) equations to predict the unmeasured ion effect (UIE) from the standard base excess (SBE), the Na−Cl effect, and the albumin effect:

$$UIE(mmol / L) = SBE - [Na - Cl - 38] - 0.25 \times [42 - albumin(g / L)] \quad (1.7)$$

This approach might explain how a decreased albumin is sometimes associated with metabolic alkalosis.

As an example, the SID can help define how changes in Cl^- concentration cause significant pH changes. If Cl^- moves across a membrane, it must either take a cation with it or exchange for another anion to maintain

TABLE 1.4 Examples of strong ion differences.

	Hypothetical solution with SID = 0	Hypothetical solution with SID = −10
Na (mmol/L)	100	100
K (mmol/L)	0	0
H^+ (nmol/L)	~0	10
Cl (mmol/L)	100	110
Lactate (mmol/L)	0	0
SID (mmol/L)	0	−10
pH	Neutral	Very acidic

electroneutrality. If Cl^- takes another strongly dissociated ion with it, such as Na, there will be no net change in the SID and no change in pH. However, if Cl^- exchanges for a weaker HCO_3 ion, the SID of the solution receiving Cl^- will decrease and the solution will become more acidic, and the solution receiving the HCO_3^- will become more alkaline *(13,14)*.

While the SID theory may correctly explain some actual ion movements and pH changes, it is rarely used clinically to understand the diagnosis and management of acid−base disorders, as noted elsewhere *(18)*.

Hemoglobin and derivatives

Hb and its derivatives are measured by cooximetry. These are oxyhemoglobin (O_2Hb), deoxyhemoglobin (HHb), carboxyhemoglobin (COHb), and methemoglobin (metHb). Briefly, O_2Hb has four oxygen molecules bound to each of the heme groups in the hemoglobin molecule. HHb is the oxygen-unloaded form of hemoglobin. COHb has carbon monoxide (CO) bound very tightly to an O_2-binding site on a heme group, which increases the affinity of the other three heme groups for O_2 molecules, and prevents oxygen unloading until the CO is released. MetHb is inactive Hb (unable to bind O_2) because its ferrous ion (Fe^{2+}) has been oxidized to a ferric ion (Fe^{3+}). Hemoglobin function is further discussed later in this chapter.

sO_2 and $\%O_2Hb$

sO_2 is the percent oxygen saturation, which is the percentage of functional Hb that is bound with oxygen. sO_2 is calculated as the concentration of O_2Hb divided by total functional Hb. sO_2 may vary from 0% to 100% *(2)*.

$$sO_2 = O_2Hb/(O_2Hb + HHb) \tag{1.8}$$

The $\%O_2Hb$ (formerly called the fractional Hb saturation) is the percentage of total Hb that is saturated with oxygen. The $\%O_2Hb$ may be used for determining the oxygen content of blood and, therefore oxygen delivery (DO_2) and oxygen consumption (VO_2) (19).

$$\%O_2Hb = O_2Hb/(O_2Hb + HHb + COHb + MetHb) \qquad (1.9)$$

The difference between sO_2 and $\%O_2Hb$ is usually of little clinical relevance, but it can be quite significant in the setting of pathologically high levels of COHb or metHb. Therefore, it is important to understand the difference between these two terms for oxygen bound to hemoglobin (sO_2 and $\%O_2Hb$) (20).

COHb and metHb

Carboxyhemoglobin (COHb) and methemoglobin (metHb) together normally make up only 1%–2% of the total Hb in blood. Neither can perform O_2 carrying and releasing functions. COHb is Hb with carbon monoxide (CO) bound very tightly to a site that normally binds O_2. Exposure to carbon monoxide will increase the %COHb from a normal baseline level of about 1% to 5%–10% in smokers and 50% or more in those with exposure to toxic or lethal levels, as can occur in the setting of combustion in a poorly ventilated area. Methemoglobin is Hb that has its Fe^{2+} ion oxidized to Fe^{3+}, which makes metHb unable to effectively bind O_2.

DO_2 and VO_2

Both DO_2 (oxygen delivery) and VO_2 (oxygen consumption) are parameters for assessing cardiac function, metabolic demands, and mitochondrial function in critically ill patients. There are now several ways of determining these parameters with laboratory measurements. DO_2 in mL/min requires measurement of the following values in arterial blood: Hb in g/L, $\%O_2Hb$ from 0 to 1.0, pO_2 in mmHg (of minor importance in this calculation), and cardiac output (CO) in L/min. The calculation is essentially determining how much oxygen is loaded on hemoglobin or dissolved in the blood and pumped to the body by the heart every minute.

$$DO_2 = CO \times (1.34 \times Hb \times \%O_2Hb + 0.03 \, pO_2) = CO \times (O_2a) \qquad (1.10)$$

where O_2a = arterial oxygen content.

VO_2 is oxygen consumption by the organs and tissues of the body. It is the difference in oxygen content between arterial (O_2a) and venous (O_2v) blood times cardiac output (CO):

$$VO_2 = CO \times (O_2a - O_2v) \qquad (1.11)$$

Physiology of acid and base production

Introduction. Most enzymatic and metabolic processes essential to life are critically dependent on pH. Thus, the hydrogen ion concentration (and pH) are tightly regulated in the human body. While only extracellular (blood) pH is measured for clinical purposes, most pH-dependent reactions occur within the cell, where the pH is slightly more acidic *(21)*. Metabolic acidosis may develop if either H^+ accumulates or HCO_3^- is lost. Whereas, metabolic alkalosis may develop from either loss of H^+ or increase of HCO_3^-. Respiratory acidosis results from increased CO_2 in the blood, while respiratory alkalosis results from decreased CO_2 in blood, with both processes largely related to effective alveolar ventilation rate.

"Metabolic acid". The vast majority (about 20 **moles** per day) of our metabolic acid is actually CO_2 produced in the mitochondria as one of the ultimate byproducts of glucose oxidation, with lesser amounts produced by metabolism of fats and amino acids. Most CO_2 combines with H_2O to produce carbonic acid, which is readily converted back to CO_2 for removal by alveolar ventilation. Thus, even though CO_2 is produced by ordinary metabolism, it is considered the "respiratory" component of our acid-base balance because alveolar ventilation directly affects the blood pCO_2.

"Lactate" acidosis: Contrary to common belief, lactic acid is virtually NEVER produced. In fact, the only reaction in the human that produces lactate is when pyruvate is converted to lactate, a reaction that actually *consumes* acid by the participation of NADH and H^+ being converted to NAD^+ *(22)*. So where does the acidosis come from during an oxygen deficit? It is produced many biochemical steps later when ATP is converted to ADP, phosphate, and hydrogen ions. Normally, oxidative phosphorylation reconverts these products back to ATP with little net change of H ions. However, without oxygen, the ADP cannot undergo oxidative phosphorylation and H ions (acid) accumulates. Lactate is independently produced from a buildup of NADH that favors the conversion of pyruvate to lactate by the enzyme lactate dehydrogenase. Thus, lactate is a marker for insufficient oxygen availability to the tissues, cells, and mitochondria, or inability of the mitochondria to extract oxygen in the setting of carbon monoxide poisoning, certain cytokines, reactive oxygen species, or other cell factors present in sepsis *(23)*.

Ketoacidosis. Ketoacidosis occurs in the setting of unregulated ketone production causing a severe accumulation of keto acids in the blood. The most common clinical presentation of ketoacidosis occurs with diabetic ketoacidosis, which develops when a person has too little insulin such that glucose cannot enter cells to generate energy. The liver begins to break down fat for energy, which produces toxic ketoacids formed by the deamination of amino acids and the breakdown of fatty acids. The two common acidic ketones produced in this process are acetoacetate and β-hydroxybutyrate. Ketoacidosis can also occur with prolonged starvation or severe alcoholism where the body had depleted carbohydrate stores and is forced to break down fat for sustenance.

Nonvolatile acids. These acids are generated as phosphates and sulfates predominantly from protein, phospholipid, and nucleic acid metabolism. Acids from these sources account for <1% of acid produced per day *(21)*.

Production of base. Generally, most "new" basic compounds come from dietary sources, with little direct metabolic production of bicarbonate or other alkaline substances. Bicarbonate is generated when CO_2 combines with H_2O and dissociates into H^+ and HCO_3^-, however, there is no net gain of an alkaline substance unless the H^+ is excreted. The kidney can retain or excrete HCO_3^- depending on the physiologic need.

Ammonia metabolism in the kidney is complex, but basically ammonia (approx 98% as NH_4^+ and 2% as NH_3 at pH 7.4) and HCO_3^- are produced in equimolar amounts by several enzymatic reactions involving glutamine (note: the term "ammonia" refers to the combination of both NH_4^+ and NH_3 in metabolism). In response to acidosis, ammonia and bicarbonate are generated. When NH_4^+ is excreted, bicarbonate remains, which is a major mechanism of bicarbonate generation in the kidney *(24)*. The proximal tubule is the main site of ammonia formation. In chronic kidney disease, the kidneys are unable to produce and excrete enough ammonia, which leads to the accumulation of acid and the development of metabolic acidosis. Bicarbonate can increase if Cl^- is lost by either vomiting or by renal loss from diuretics. Generally, any loss of acid will increase the proportion of HCO_3^- from CO_2. Bicarbonate given in excess to treat an acidosis is an iatrogenic source of base.

Buffer systems

Bicarbonate–carbon dioxide (carbonic acid). Bicarbonate is the base in the highest concentration ($\sim$24 mmol/L) in the blood plasma and is of central importance in acid buffering in the blood. CO_2 is a volatile acidic gas that is soluble in water (solubility factor 0.03 mmol/mmHg) and is the major acid produced as a product of energy metabolism. CO_2 produced by metabolism readily diffuses from cells into the blood, where it combines with H_2O to produce carbonic acid, which immediately dissociates to HCO_3^- and hydrogen ions.

pH, HCO_3^-, and pCO_2 (in mmHg) are related by the equation:

$$pH = pK + \log\left[HCO_3^- / (0.03 \times pCO_2)\right] \qquad (1.12)$$

Note that the pK is defined as the pH at which the HCO_3^- and H_2CO_3 (or $0.03 \times pCO_2$) are in equal concentrations. That is, their ratio is 1, and the log of 1 is equal to 0.

At the normal concentration ratio in the blood of 20:1 ("ideal" would be 1:1), and with a pK of 6.1 ("ideal" would be 7.4), the HCO_3^- - H_2CO_3 buffer system would seem to be ill-suited to buffering pH in the blood. However, this excess of base (HCO_3^-), along with the volatility of CO_2, gives tremendous ability to prevent excess accumulation of acid. The lungs effectively make this an open system, with the ability to exhale CO_2 providing almost unlimited buffering capacity. As the previous equation indicates, the pH of blood is largely determined by the ratio of HCO_3^- to H_2CO_3. Thus, $HCO_3^-:H_2CO_3$ in a concentration ratio of 15:0.75 has the same pH as a ratio of 20:1.0, but would have less buffering capacity.

Hemoglobin. Hb acts as a buffer by transporting acid from the tissues to the lungs. A remarkable feature of Hb is that it increases its affinity for hydrogen ions (H^+) as it releases oxygen. That is, deoxy-Hb (also called reduced Hb) has a greater affinity for H^+ than does O_2Hb. As oxy-Hb enters tissue capillaries, the environment of lower pO_2 and higher H^+ and CO_2 increases H^+ binding to Hb, thus promoting more release of O_2 to the tissues. As the deoxyHb enters alveolar circulation and O_2 binds to Hb, it releases H^+, that combines with HCO_3^- to ultimately form CO_2 that diffuses into the alveoli and is exhaled. 2,3-Diphosphoglycerate (2,3-DPG) also affects the Hb binding of O_2 as it binds to Hb in the tissues (decreasing affinity of Hb for O_2) and releases from Hb in the alveoli (increasing affinity of Hb for O_2). These relationships are shown in Figs. 1.2 and 1.3.

Phosphate. The $HPO_4^{2-} -H_2PO_4^-$ buffer pair is of minor importance as a buffer in plasma, with a concentration of $\sim$ 1 mmol/L (3.1 mg/dL). It is of greater importance, and in higher concentration, as an intracellular buffer.

Albumin and other proteins. Largely due to the imidazole groups on the amino acid histidine, with a pK of $\sim$ 7.4, albumin and other proteins also act as pH buffers. Albumin is the major "unmeasured" anion in blood, and it is normally the biggest contributor to the anion gap. Patients with critical illness and/or malnutrition commonly have hypoalbuminemia, which can lower the AG and potentially impact the interpretation of this parameter *(11,25)*.

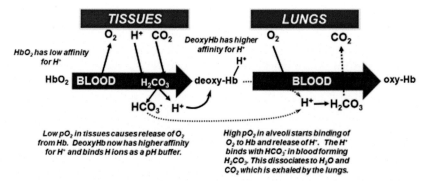

FIGURE 1.2   Interrelationships of the bicarbonate and hemoglobin (Hb) systems in buffering H^+ and in the binding and release of oxygen in lungs and tissues. The high pO_2 in the lungs and low pO_2 in the capillaries, respectively, drive the binding and release of O_2. H ions participate by binding to Hb in the tissue capillaries to decrease the affinity of Hb for O_2 or dissociating from Hb in the lungs to increase the affinity of Hb for O_2.

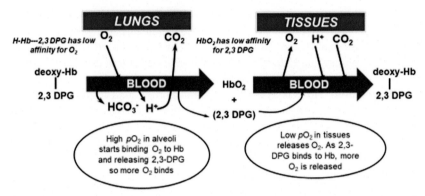

FIGURE 1.3   Interrelationships of hemoglobin (Hb), 2,3-diphosphoglycerate (2,3-DPG), and H^+ in binding and release of oxygen in lungs and tissues. 2,3-DPG remains in red cells and cooperates with Hb to increase or decrease Hb affinity for O_2. As deoxy-Hb bound with 2,3-DPG enters alveolar circulation in the lungs, O_2 binds to Hb, which releases 2,3-DPG and H^+. These further increase the affinity of Hb for O_2 so more O_2 is bound. As O_2-Hb enters tissues, the environment of higher H^+ and CO_2 promotes the release of O_2 to the tissues, which now favors Hb binding of 2,3-DPG and H^+ to promote more release of O_2 from Hb.

Acid–base regulation

Normal acid–base balance, oxygen metabolism, and their associated disorders can involve a complex interplay of several organ systems, especially pulmonary, renal, hepatic, and gastrointestinal. The brain may be overlooked but it plays an extremely important role in acid–base regulation *(21)*.

Respiratory (pulmonary) system. Because arterial CO_2 is influenced greatly by alveolar ventilation, pCO_2 is considered the respiratory component of the bicarbonate-CO_2 buffer system. Because CO_2 is the end product of many aerobic metabolic processes, buffering and removal of CO_2 are continually required for pH regulation. Provided there is a sufficient gradient of pCO_2 between tissues and blood, CO_2 will readily diffuse into the blood. As mentioned previously, CO_2 combines enzymatically with H_2O to form the unstable H_2CO_3, which quickly dissociates into HCO_3^- and H^+ ions. Deoxy-Hb plays a key role here by readily accepting the H^+ and transporting it to the lungs from the tissues. As H^+ is exchanged for oxygen in the lungs, H^+ quickly combines with HCO_3^- to produce H_2CO_3, which dissociates to H_2O and dissolved CO_2, which diffuses into the alveolar air for ventilatory removal (see Fig. 1.2).

The arterial pCO_2 represents a balance between tissue production and diffusion of CO_2 into the blood and pulmonary removal of CO_2. An elevated pCO_2 usually indicates inadequate ventilation (hypoventilation) and a respiratory acidosis. Conversely, a decreased pCO_2 usually indicates excessive ventilation (hyperventilation) and a respiratory alkalosis. There are several causes of respiratory abnormalities *(26,27)*.

Respiratory acidosis (ventilatory failure) can be caused by the following:

- Obstructive lung disease: chronic bronchitis, emphysema, asthma
- Impaired central respiratory drive: head trauma, sedatives, opioids, anesthesia
- Weak or disordered diaphragm muscles
- Hypoventilation by mechanical ventilator

Respiratory alkalosis (hyperventilation) can be caused by the following:

- Hypoxemia (which stimulates hyperventilation)
- Anxiety
- Pain
- Hyperventilation by mechanical ventilator
- Metabolic acidosis
- Septicemia

Metabolic (renal) system. When the H^+ concentration deviates from normal, the kidneys respond by reabsorbing or secreting hydrogen, HCO_3^-, NH_4^+, and other ions to regulate the pH of the blood. Because HCO_3^- is primarily regulated by the kidney, it is considered the metabolic component of the HCO_3^--CO_2 buffer system. In kidney failure, acidemia becomes more likely because the kidneys are less able to generate ammonia and bicarbonate so that less NH_4^+ is excreted which leaves less bicarbonate to buffer the hydrogen ion.

Metabolic (liver) system. The urea cycle in the liver essentially converts bicarbonate and ammonia to urea, which is a neutral compound. Because acidosis decreases the hepatic production of urea, more ammonia is converted to glutamine, which spares bicarbonate and attenuates the acidosis *(21)*.

Brain system. The brain may be an overlooked but very important region for acid–base regulation *(21)*. Regulation of brain pH differs from peripheral pH regulation by involving cellular mechanisms that allow the brain to compensate back to near normal in each of the primary acid–base disturbances: metabolic acidosis and alkalosis, and respiratory acidosis and alkalosis. Correction of systemic pH when brain pH is disturbed, such as in stroke, may disrupt the important brain compensation *(21)*.

Aldosterone in acid–base balance. Aldosterone is a very powerful regulator of acid–base balance and links this with its principal function in regulating salt and potassium homeostasis *(28)*.

Compensation. Compensation is a homeostatic response to an acid–base disorder in which the body attempts to restore pH toward normal by adjusting the HCO_3^-:$(0.03 \times pCO_2)$ back to a normal ratio of 20:1. Compensation involves either a relatively rapid ventilatory response (change in pCO_2) to a metabolic abnormality or a relatively slow metabolic response (change in HCO_3^-) to a ventilatory abnormality. Although a change in alveolar ventilation can alter arterial pH in minutes, the kidneys require hours to days to significantly affect pH by altering the excretion of HCO_3^-.

As compensation returns pH toward normal, the pH-driven compensation process slows, then stops before the pH is fully normalized, which has survival benefits. Mechanisms for this are complex, but in metabolic acidosis, the acid pH and low bicarbonate stimulate the medullary center to increase ventilation. Because hyperventilation increases energy consumption, incomplete compensation returns pH sufficiently near normal while saving energy to maintain other vital processes *(21)*.

In metabolic alkalosis, hypoventilation occurs to decrease the pH. However, full compensation is limited because coexisting hypokalemia, volume depletion, and hyperaldosteronism develop and maintain a slight metabolic alkalosis. The benefits of this incomplete compensation are that water, salt, and potassium balance are maintained.

While the respiratory compensation by hyperventilation is fairly predictable in metabolic acidosis, the respiratory response by hypoventilation is less predictable in metabolic alkalosis. While hypoventilation almost invariably occurs in metabolic alkalosis, other factors such as pain and hypoxemia can stimulate ventilation and overcome the hypoventilatory effect of metabolic alkalosis *(29)*.

The expected compensation for each acid—base abnormality will be discussed later in the sections "Clinical Disorders of Acid-Base Balance" and "Mixed acid-base disorders."

Hemoglobin function

Hb is a protein of 64,500 Da that consists of four heme molecules attached to four globin molecules. Hb is certainly a hall-of-fame molecule, having the essential abilities to bind, transport, and release oxygen to the tissues and to transport H^+ and carbon dioxide from the tissues to the lungs. Each of the four heme groups contains an Fe^{++} ion and can bind one molecule of oxygen. Structural changes occur with oxygen binding that results in color changes to the molecule. HHb (deoxyhemoglobin) gives the blood a deep purplish hue, while O_2Hb gives the blood a scarlet red appearance *(30)*. 2,3-DPG is a molecule contained within erythrocytes that cooperates with Hb to bind or release O_2. As Hb binds O_2 in the alveoli, this favors the release of 2,3-DPG, which further increases the affinity of Hb for O_2 binding (Fig. 1.3). This eventually leads to the saturation of Hb with oxygen and a sigmoidal relationship between pO_2 and sO_2, as shown in the Hb-oxygen dissociation curve in Fig. 1.4. Hemoglobin will be discussed in detail in Chapter 4.

Reference intervals for blood gases

Higgins provides his usual excellent discussion of the importance and derivation of reference intervals, specifically for blood gases *(31,32)*. Because the clinical value of laboratory test results depends on the quality of reference

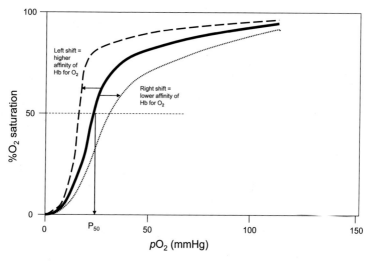

FIGURE 1.4 Oxyhemoglobin dissociation curve for whole blood. The solid middle curve represents the percentage of Hb that is saturated with oxygen as pO_2 is increased. The dashed curve represents a *left shift* of the curve when the affinity of Hb for O_2 is increased by cooler temperature, higher pH, or release of 2,3-DPG. The dotted curve represents a *right shift* of the curve when the affinity of Hb for O_2 is decreased by warmer temperature, lower pH, or binding of 2,3-DPG. The P50 is the oxygen pressure (pO_2) that saturates Hb by 50% and is an index of the overall affinity of Hb for O_2. A *left shift* of the curve decreases the P50 and a *right shift* increases the P50.

intervals, they should be reviewed and updated regularly. He also notes the lack of good studies of reference intervals for arterial blood gases, understandably due to the risk and pain associated with collecting arterial blood. Higgins cites a study from Denmark done in 2011 in which arterial blood was collected from 182 healthy adults (96 females; 86 males) ages 20–76 years (*33*). Of critical importance to blood gas results, all samples were analyzed within 10 min of collection. Reference intervals from this study are included in Table 1.1 along with reference intervals from several sources.

The reference intervals for pO_2 require additional explanation. First, because pO_2 has some variability with age, enough data is provided to indicate the pO_2 differences for various age groups. After age 60 years, the average pO_2 declines, although there is debate about the rate at which this occurs. Traditionally, the pO_2 is believed to decline at about 1 mmHg per year after age 60, but this has been challenged (*34,35*). While a lower reference limit for pO_2 of 70 mmHg is reasonable at 70 years of age, that may also be the lower limit for persons older than 70 years who are in reasonable health. Second, some claim that for persons breathing room air, the arterial pO_2 should not go above 100 mmHg. This is based on the presumption that alveolar pCO_2 is approximately 50 mmHg for a venous pCO_2 of 40 mmHg. Because alveolar air is a mixture of incoming atmospheric air (pO_2 ~ 150 mmHg) and with the CO_2

contribution to alveolar air of $\sim pCO_2$ 50 mmHg, the alveolar pO_2 should be no higher than ~ 100 mmHg. However, in younger persons, the studies clearly indicate the arterial pO_2 can be up to ~ 110 mmHg (*31,33*).

In the neonate, Brouillette and Waxman showed that blood gases change rapidly during the first 60 min after birth, then stabilize by 60–120 min after birth at levels that resemble adult values (*36*).

Self-assessment and mastery

1. Who is credited with discovering the concept of pulse oximetry?
   a. Leland Clarke
   b. Peter Stewart
   c. Takuo Aoyagi
   d. John Toffaletti
   e. Van Slyke
2. Which parameter is most closely related to base excess?
   a. Anion gap
   b. Na
   c. Cl
   d. HCO_3
3. Which anions contribute to the anion gap?
   a. Albumin
   b. Acetoacetate
   c. Phosphate
   d. Lactate
   e. All the above
4. Which molecule binds most tightly to hemoglobin?
   a. Oxygen
   b. Hydrogen ion
   c. CO
   d. CO_2
5. What is VO_2?
   a. Vanadium dioxide
   b. Breathing rate per minute
   c. Oxygen consumption
   d. Oxygen delivery
   e. Percent of oxygen in inspired air
6. Which statement is false?
   a. Normal metabolism produces more alkaline than acidic compounds
   b. Loss of HCO_3 creates an acidosis
   c. Loss of H ion creates and alkalosis
   d. Extracellular pH is higher than the intracellular pH

7. Which metabolic product creates the most acidic condition?
   a. CO_2
   b. NAD+
   c. ATP
   d. Lactate
8. Most ammonia is produced as
   a. NH_4^+
   b. Amino acids
   c. NH_3
   d. Creatinine
9. What are functions of hemoglobin? Choose all that apply.
   a. Buffers pH
   b. Carries O_2
   c. Releases O_2
   d. Uses 2,3-DPG in O_2 binding and release
10. Which organs contribute to acid–base regulation?
    a. Lungs
    b. Kidneys
    c. Liver
    d. Brain
    e. All the above
11. Compensation in acid–base regulation responds to which of the following parameters in blood?
    a. Changes in lactate
    b. Changes in pH
    c. Changes in sodium
    d. Changes in creatinine
12. Arterial pCO_2 is a balance between:
    a. CO_2 production and binding by Hb
    b. Tissue production and pulmonary removal of CO_2 from blood
    c. HCO3 retention and excretion by the kidneys
    d. Hypoventilation and hyperventilation
    e. Lactate production and ATP depletion

Answer Key:

1. c
2. d
3. e
4. c
5. c
6. a
7. a

8. a
9. a, b, c, d
10. e
11. b
12. b

References

1. Grogono, A. W. *Acid-Base Tutorial: Acid-Base History.* http:/www.acid-base.com/history. (accessed December 2020).
2. Story, D. A. Bench-to-Bedside Review: A Brief History of Clinical Acid-Base. *Crit. Care* **2004,** *8,* 253–258.
3. Astrup, A.; Severinghaus, J. W. *The History of Blood Gases, Acids and Bases;* Munksgaard: Copenhagen, 1986.
4. Grogono, A. W. *Acid-Base Tutorial: Terminology: Base Excess.* http:/www.acid-base.com. (accessed December 2020).
5. Kofstad, J. *All About Base Excess — to BE or Not to BE,* July 2003. www.acutecaretesting. org.
6. *Base Excess & Calculated Bicarbonate.* http://www-users.med.cornell.edu/~spon/picu/calc/ basecalc.htm. (accessed October 2020).
7. Connelly, C. R.; Schreiber, M. A. Endpoints in resuscitation. *Curr. Opin. Crit. Care* **2015,** *21*(6), 512–519.
8. Ibrahim, I.; Chor, W. P.; Chue, K. M.; et al. Is arterial base deficit still a useful prognostic marker in trauma? A systematic review. *Am. J. Emerg. Med.* **2016,** *34*(3), 626–635.
9. Kraut, J. A.; Nagami, G. T. The Serum Anion Gap in the Evaluation of Acid-Base Disorders: What Are Its Limitations and Can Its Effectiveness Be Improved? *Clin. J. Am. Soc. Nephrol.* **2013,** *8,* 2018–2024.
10. Feldman, M.; Soni, N.; Dickson, B. Influence of Hypoalbuminemia or Hyperalbuminemia on the Serum Anion Gap. *J. Lab. Clin. Med.* **2005,** *146,* 317–320.
11. Brandis, K. *Acid-Base Physiology: The Anion Gap.* From http://anaesthesiaMCQ.com/ AcidBaseBook/ab3_2. (accessed December 2020).
12. Iberti, T. J.; Leibowitz, A. B.; Papadakos, P. J.; Fischer, E. P. Low Sensitivity of the Anion Gap as a Screen to Detect Hyperlactatemia in Critically Ill Patients. *Crit. Care Med.* **1990,** *18,* 275–277.
13. Story, D. A.; Morimatsu, H.; Bellomo, R. Strong Ions, Weak Acids and Base Excess: A Simplified Fencl-Stewart Approach to Clinical Acid-Base Disorders. *Br. J. Anaesth.* **2004,** *92,* 54–60.
14. Grogono, A. W. *Acid-Base Tutorial: Stewart's Strong Ion Difference.* http:/www.acid-base. com/strong-ion-difference. (accessed December 2020).
15. Moviat, M.; Terpstra, A. M.; Ruitenbeek, W.; Kluijtmans, L. A. J.; Pickkers, P.; van der Hoeven, J. G. Contribution of Various Metabolites to the "Unmeasured" Anions in Critically Ill Patients with Metabolic Acidosis. *Crit. Care Med.* **2008,** *36* (3), 752–758.
16. *ICU Calculator; Blood Gas (Stewart).* https://intensivecarenetwork.com/Calculators/Files/ Gazo.html.
17. Lloyd, P. Strong Ion Calculator — A Practical Bedside Application of Modern Quantitative Acid-Base Physiology. *Crit. Care Resusc.* **2004,** *6,* 285–294.

18. Masevicius, F. D.; Dubin, A. Has Stewart Approach Improved Our Ability to Diagnose Acid-Base Disorders in Critically Ill Patients? *World J. Crit. Care Med.* **2015,** *4* (1), 62−70.

19. Scott, M. G.; LeGrys, V. A.; Hood, J. L. Electrolytes and Blood Gases. In *Tietz Textbook of Clinical Chemistry and Molecular Diagnostics;* Burtis, C. A., Ashwood, E. R., Bruns, D. E., Eds., 5th ed.; Elsevier Saunders: St Louis, 2012; pp 807−835.

20. Toffaletti, J.; Zijlstra, W. G. Misconceptions in Reporting Oxygen Saturation. *Anesth. Analg.* **2007,** *105,* S5−S9.

21. Seifter, J. L.; Chang, H.-Y. Disorders of Acid-Base Balance: A New Perspective. *Kidney Dis. (Basel)* **2017,** *2* (4), 170−186.

22. Robergs, R. A.; Ghiasvand, F.; Parker, D. Biochemistry of Exercise-Induced Metabolic Acidosis. *Am. J. Physiol. Regul. Integr. Comp. Physiol.* **2004,** *287,* R502−R516.

23. Gotts, J. E.; Matthay, M. A. Sepsis: Pathology and Clinical Management. *Br. Med. J.* **2016,** *353,* i1585.

24. Mohiuddin, S. S.; Khattar, D. *Biochemistry, Ammonia;* NCBI Bookshelf, July 2020.

25. Kraut, J. A.; Madias, N. E. Serum Anion Gap: Its Uses and Limitations in Clinical Medicine. *Clin. J. Am. Soc. Nephrol.* **2007,** *2,* 162−174.

26. Patel, S.; Sharma, S. *Respiratory Acidosis;* NCBI Bookshelf, June 2020. www.ncbi.nlm.nih.gob/books/NBK482430/.

27. Brandis, K. *Respiratory Alkalosis − Causes.* http://www.anaesthesiaMCQ.com. (accessed November 2020).

28. Wagner, C. A. Effect of Mineralocorticoids on Acid-Base Balance. *Nephron. Physiol.* **2014,** *128,* 26−34.

29. Brandis, K. *Acid-Base physiology: Metabolic Alkalosis - Compensation.* From http://www.anaesthesiaMCQ.com/AcidBaseBook/ab7_5. (accessed January 2020).

30. *Why/How is Blood Red? (Colours of Hemoglobin).* https://chemistry.stackexchange.com/questions/49940/why-how-is-blood-red-colours-of-hemoglobin.

31. Higgins, C. *Adult Reference Intervals for Blood Gases,* January 2012. www.acutecaretesting.org.

32. Higgins, C. *Central Venous Blood Gas Analysis,* July 2011. www.acutecaretesting.org.

33. Klaestrup, E.; Trydal, T.; Pedersen, J.; Larsen, J.; Lundbye-Christensen, S.; Kristensen, S. Reference Intervals and Age and Gender Dependency for Arterial Blood Gases and Electrolytes in Adults. *Clin. Chem. Lab. Med.* **2011,** *49,* 1495−1500.

34. Hardie, J. A.; Vollmer, W. M.; Buist, S.; Ellingsen, I.; Morkve, O. Reference Values for Arterial Blood Gases in the Elderly. *Chest* **2004,** *125,* 2053−2060.

35. Sorenson, H. M. Arterial Oxygenation in the Elderly. *Elite Learn.* **2006,** *19* (2), 17. https://www.elitecme.com/resource-center/respiratory-care-sleep-medicine/arterial-oxygenation-in-the-elderly.

36. Brouillette, R. T.; Waxman, D. H. Evaluation of the Newborn's Blood Gas Status. *Clin. Chem.* **1997,** *43* (1), 215−221.

Chapter 2

Physiologic mechanisms and diagnostic approach to acid—base disorders

Metabolic acidosis

Metabolic acidosis is defined as a clinical process that leads to a decreased blood pH (acidemia) and a decreased HCO_3 level. It is caused by a gain of acidic compounds and/or loss of a base, usually bicarbonate. Electrolyte movements can also play a role in acid—base balance that are sometimes manifested as an increased anion gap.

Physiologic mechanisms of metabolic acidosis *(1—3)*:

- In diarrhea, the pancreatic secretions into the small intestines have a relatively high sodium to chloride content, with a relatively higher bicarbonate concentration. Loss of such GI fluids often leads to a metabolic acidosis.
- In renal tubular acidosis, if the excretion of sodium and potassium are in excess of chloride excretion into the urine, the anion accompanying sodium/potassium is usually bicarbonate. This can lead to a decreased serum bicarbonate and metabolic acidosis.
- Hyperchloremia usually indicates a metabolic acidosis. Gains of chloride by GI absorption or intravenous administration can cause hyperchloremia. This is a physiologically complex area, but here are a few principles:
  - If the Cl^- concentration increases and the Na^+ concentration stays the same, an anion other than Cl^- must decrease. This anion is usually bicarbonate.
  - Hyperchloremia can occur with loss of nonchloride sodium salts from GI tract or from urine.
  - Hyperchloremia will develop with additions of fluids containing chloride salts such as sodium, potassium, calcium, or ammonium.
  - Gains or losses of potassium can also affect the sodium concentration, which will then affect the chloride concentration.

Blood Gases and Critical Care Testing. https://doi.org/10.1016/B978-0-323-89971-0.00001-X

An elevated anion gap can be associated with a metabolic acidosis. The following processes are usually associated with an increased anion gap:

- Lactate and acid production from anaerobic metabolism. If adequate oxygen is not available to tissues (hypoxia), cellular conditions begin to favor anaerobic metabolism with the conversion of pyruvate to lactate. While clinicians often think of this process producing lactic acid, it is actually lactate that is produced (4). The acid (H^+) comes from the degradation of large amounts of ATP without resynthesis.
- Excess production of acid from ketoacidosis. The most common type of ketoacidosis is diabetic ketoacidosis, which occurs when glucose cannot enter cells to produce adequate ATP. In the setting of low cellular glucose (serum glucose is often quite elevated in the setting of diabetic ketoacidosis), fatty acids will be oxidized to produce ketoacids for energy. Ketoacidosis can also be seen with starvation where plasma glucose levels are low, and in patients who are malnourished and suffer from chronic alcoholism.
- Ingestion of exogenous acids or acid-producing substances: salicylates, ethanol, ethylene glycol, methanol, etc.
- Inadequate renal excretion of acids. In renal failure, depending on whether glomerular or tubular failure, accumulation of H^+ and/or loss of HCO_3^- may occur. Because renal acidosis usually has a slow onset, hyperventilation typically provides respiratory compensation to prevent acidemia in the early stages.

Use of the anion gap in metabolic acidosis. The anion gap (AG), especially when elevated, may be useful in diagnosing the type of metabolic acidosis and in indicating the possibility of a mixed acid–base disorder (5), as shown in Table 2.1.

As described earlier, the AG is the difference between the commonly measured cations (Na^+ and sometimes K^+) and the commonly measured

TABLE 2.1 Changes of anion gap in various acid–base disorders.

Disorder	Decreased	Gained	Effect on AG
Diarrhea	HCO_3^-	Cl	Little
Renal tubular acidosis	HCO_3^-	Cl	Little
Lactate acidosis	HCO_3^-	Lactate	Increased
Ketoacidosis	HCO_3^-	Ketoacids	Increased
Mixed disorder: ketoacidosis with metabolic alkalosis	HCO_3^-	Ketoacids and	Increased (with little change HCO_3^- in total CO_2)

anions (Cl^- and HCO_3^-). The approximate reference intervals for the AG are 8–16 mmol/L (8–16 mEq/L) when the $AG = Na-(Cl + CO_2)$ and 12–20 mmol/L (12–20 mEq/L) when $AG = (Na + K)-(Cl + CO_2)$.

Serum albumin as an important factor affecting the anion gap calculation. Because albumin in blood constitutes a large proportion of the "unmeasured" anions, a decrease or increase (less common) in the albumin concentration will, respectively, decrease or increase the anion gap calculation. For each g/L decrease in albumin concentration, the AG is decreased by approximately 0.25 mmol/L (5,6). If the lab reports albumin in g/dL then each g/dL decrease in albumin concentration will result in the AG decreasing by approximately 2.5 mmol/L. This could hide an increased AG due to increased lactate in a metabolic acidosis. If hypoalbuminemia is noted, which is common in critically ill patients, the AG can be "corrected" using Eq. (2.1) to better interpret the AG. The sensitivity of an increased anion gap in detecting hyperlactatemia appears to be improved by correcting the AG for the albumin concentration (5).

$$AG\ corr = AG\ meas + 0.23 \times (45 - Alb\ meas) \tag{2.1}$$

AG corr = corrected AG; AG meas = measured AG; Alb meas = measured albumin in g/L.

In Table 2.1, the AG measured (really calculated) on a hypothetical sample with an elevated lactate of 5.0 mmol/L and a normal albumin of 45 g/L is elevated at 18 mmol/L, as expected. However, on a sample with the same elevated lactate of 5.0 mmol/L, but a decreased albumin of 25 g/L, the measured AG is apparently normal at 12 mmol/L. Correcting the AG in this latter sample gives a corrected AG of 18 mmol/L. Table 2.2 also shows a

TABLE 2.2 Effects of albumin concentration on the anion gap.

Chemistry test	Normal albumin Elevated lactate	Low albumin Elevate lactate	Elevated albumin Normal lactate
Na (mmol/L)	140	138	140
K (mmol/L)	4.0	4.0	4.0
Cl (mmol/L)	104	108	98
HCO_3 (mmol/L)	18	18	24
Albumin (g/L)	45	25	70
Lactate (mmol/L)	5.0	5.0	2.0
AG meas	18	12	18
AG corr	18	18	12

hypothetical sample with an elevated albumin of 70 g/L and a normal lactate that had a spuriously increased AG of 18 mmol/L. Correcting the AG for the increased albumin gave a normal AG of 12 mmol/L. A potential concern is that this correction requires using a "normal" albumin in the calculation, such as the mean of the reference interval. It would probably be better to use the patient's baseline albumin and AG as a reference if available.

Monitoring AG results in the laboratory is also a common quality control technique for detecting drift in electrolyte results. Note that this is detected by a series of unusual anion gaps, not by a single result. This could be why one report claims that the most common cause of low anion gaps is laboratory error in electrolyte measurements (7).

Expected compensation in metabolic acidosis (8,9) (Table 2.3 and Fig. 2.1). In the initial phase of metabolic acidosis (acidemia with a low plasma bicarbonate), both pH and HCO_3^- are decreased while pCO_2 remains normal. The expected respiratory compensation is hyperventilation, which will begin to lower blood pCO_2 within 30 min, have a significant effect by 2 h, and reach maximal effect in 12−24 h (9). A general rule is that respiratory compensation lowers pCO_2 by 1.2 mmHg for each 1.0 mmol/L decrease in HCO_3^- below 24 mmol/L (10). For example, in a patient who develops metabolic acidosis and has an initial pCO_2 of 40 mmHg and initial HCO_3^- of 24 mmol/L, if the HCO_3^- decreases by 8 mmol/L from 24 to 16 mmol/L, the pCO_2 after compensation should be approximately 30 mmHg (40−1.2 × 8). The equation for this relationship is:

$$\text{Compensated } pCO_2 = \text{Initial } pCO_2 - 1.2$$

$$\times \left[\text{Initial } HCO_3^- - \text{Compensated } HCO_3^- \right] \quad (2.2)$$

$$pCO_2 = 40 - 1.2 \times [24 - HCO_3] = 40 - 10 = 30 \text{ mmHg}$$

Thus, if the pCO_2 is more than 3 mmHg above this calculated value, the patient may not be fully capable of compensating through hyperventilating and thus may also have an underlying respiratory acidosis. If the pCO_2 is more than 3 mmHg below this calculated value, an underlying respiratory alkalosis may be present. Either of these situations would indicate that a mixed disorder is present (metabolic acidosis with either respiratory acidosis or respiratory alkalosis).

Caution: These calculations for expected compensation should be used as guidelines in the diagnosis and should not be overinterpreted as an absolute diagnosis.

Treatment. The optimal treatment for metabolic acidosis is correcting the underlying cause, such as by administering insulin for diabetic ketoacidosis, improving oxygen delivery by increasing the oxygen content of the blood and/or improving cardiac output. Losses of fluid and electrolytes should be replaced if appropriate, dialysis for renal failure or to remove certain

TABLE 2.3 pH, HCO$_3^-$, and pCO$_2$ in primary disorders, compensation, and mixed disorders.

Primary disorder	pH	HCO$_3^-$	pCO$_2$	Condition
Metabolic acidosis	Dec	Dec	Norm	Early phase of met acid (<2 h)
	Dec to norm	Dec	Dec	Expected Resp compensation (≥6 h)
	Dec	Dec	Norm to Dec	Inability to compensate
	Dec++	Dec	Inc	Mixed met acid and Resp acid
Metabolic alkalosis	Inc	Inc	Norm	Early phase of met acid (<2 h)
	Inc to norm	Inc	Inc	Expected Resp compensation (≥6 h)
	Inc	Inc	Norm to Inc	Inability to compensate
	Inc++	Inc	Dec	Mixed met Alk and Resp Alk
Respiratory acidosis	Dec	Norm	Inc	Initial phase of Resp acid
	Dec to norm	Inc	Inc	Expected met compensation (≥24 h)
	Dec	Norm to inc	Inc	Inability to compensate
	Dec++	Dec	Inc	Mixed Resp acid and met acid
Respiratory alkalosis	Inc	Norm	Dec	Initial phase of Resp Alk
	Inc to norm	Dec	Dec	Expected met compensation (≥24 h)
	Inc	Norm to Dec	Dec	Inability to compensate
	Inc++	Inc	Dec	Mixed Resp Alk and met Alk

++, markedly; *Dec*, decreased; *Inc*, increased; *met Acid*, metabolic acidosis; *met Alk*, metabolic alkalosis; *Met*, metabolic; *Norm*, normal; *Resp Acid*, respiratory acidosis; *Resp Alk*, respiratory alkalosis; *Resp*, respiratory.

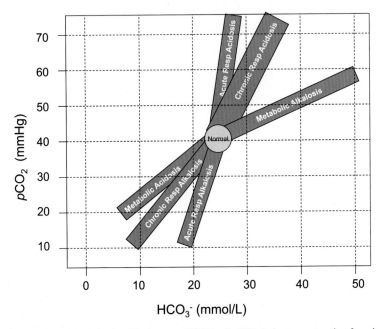

FIGURE 2.1 Expected relationships between HCO_3^- and pCO_2 during compensation for primary metabolic acidosis and alkalosis, and for acute and chronic phases of respiratory acidosis and alkalosis. The areas shown are based on plots of Eqs. (2.2–2.4, 2.6, 2.8, 2.9).

exogenous causes of metabolic acidosis and possible administration of bicarbonate. However, administration of bicarbonate may be harmful and is not generally recommended outside of hemodynamic instability secondary to profound metabolic acidosis *(11)*.

Metabolic alkalosis

Metabolic alkalosis is an acid–base disorder characterized by an elevation in bicarbonate level above 26 mmol/L and an elevation in pH. Many disease states can induce metabolic alkalosis including excess renal loss of hydrogen ions or retention of bicarbonate, intracellular shifts of hydrogen ions, GI loss of hydrogen ions, and fluid volume loss through diuretic use without proportional loss of bicarbonate, known as "contraction alkalosis" *(12)*, which is frequently associated with renal impairment because a healthy kidney can excrete large amounts of HCO_3 when it is in excess *(13)*.

Mechanisms of metabolic alkalosis:

- Increased mineralocorticoid levels. The kidney uses an exchange mechanism in which sodium is reabsorbed and hydrogen is excreted. Because aldosterone increases retention of sodium, conditions that increase aldosterone, such as primary aldosteronism (Conn syndrome) and diuretics, can lead to hypernatremia and increased loss of hydrogen ions in the urine *(12)*. Because cortisol and other steroids such as corticosterone have some mineralocorticoid activity, excess cortisol, as in Cushing's syndrome, can lead to excretion of both K^+ and H^+ in the distal renal tubules with subsequent metabolic alkalosis and hypokalemia.

- GI loss of acid by vomiting. The stomach fluids are highly acidic, with a pH of approximately 1.5−3.5. Thus, the loss of gastric acid causes metabolic alkalosis with an increased bicarbonate concentration in the blood.
- Hypochloremia. Deficiency of Cl^- in blood enhances HCO_3^- reabsorption in the renal tubule. When cations such as Na^+ and K^+ are reabsorbed, an anion must follow. Because less Cl^- is available, more HCO_3^- is reabsorbed.
- Hypokalemia. Hypokalemia stimulates distal tubular reabsorption of HCO_3 ions. In primary mineralocorticoid excess, aldosterone increases tubular Na reabsorption and promotes the loss of K and H ions. These movements lead to extracellular alkalosis with hypochloremia, hypokalemia, and expanded ECF volume.
- Administration of excess HCO_3^-. Although not usually required for treatment of metabolic acidosis, infusion of sodium bicarbonate solution may sometimes be used. If excessive, HCO_3^- administration may lead to alkalosis, especially if renal function is compromised.
- Contraction alkalosis. This occurs when a large volume of fluid with relatively more sodium than bicarbonate is lost from the body such that the plasma bicarbonate concentration increases. This can occur with diuretics, diarrhea, cystic fibrosis, and other conditions *(12)*.
- Hypoalbuminemia. According to the strong ion difference concept *(13)*, a decrease in plasma albumin (being the predominant weak acid in plasma) causes a decreased weak acid concentration that results in a metabolic alkalosis.

Expected compensation in metabolic alkalosis (Table 2.3 and Fig. 2.1). Because the normally functioning kidney can excrete large amounts of bicarbonate, a moderate bicarbonate load can be readily excreted to correct a metabolic alkalosis. However, if renal bicarbonate excretion is hindered, bicarbonate can increase in the blood leading to an increase in blood pH.

Changes in blood pH are sensed by both peripheral and cerebral chemoreceptors, such that a metabolic alkalosis consistently causes hypoventilation and an increase in blood $p\mathrm{CO}_2$ to compensate the increased bicarbonate. However, while hypoventilation almost invariably occurs when metabolic alkalosis is present, other factors such as pain, for example by the arterial puncture, or hypoxemia may stimulate hyperventilation and overcome the hypoventilatory stimulus of metabolic alkalosis *(14,15)*.

Because they may partially offset each other, the combined stimuli for hypoventilation by alkalemia and hyperventilation by hypoxemia can present both diagnostic and therapeutic challenges. Patients who present with hypercapnia from metabolic alkalosis and hypoxemia from other causes, may have the metabolic alkalosis missed. In such cases, administering oxygen should decrease the hypoxemic stimulus for hyperventilation and reveal the hypoventilation as a compensatory response to metabolic alkalosis (15).

Although the maximal respiratory compensation by hypoventilation was once believed to be no more than 55–60 mmHg, compensation can elevate arterial pCO_2 to over 80 mmHg in some instances (14,15).

In the initial phase of metabolic alkalosis, both pH and HCO_3^- are increased and pCO_2 remains normal. Within 2 h, compensation by hypoventilation should be apparent, with maximal compensation occurring by 12–24 h, which increases the blood pCO_2 to match the elevated HCO_3^- and restore the HCO_3^-/pCO_2 ratio to normal. In general, respiratory compensation increases pCO_2 by 0.7 mmHg for each 1.0 mmol/L rise in HCO_3^-. For example, in a patient who develops metabolic alkalosis, after 24 h, if the HCO_3^- increases from 24 to 32 mmol/L (8 mmol/L above a normal 24 mmol/L), the pCO_2 should be approximately 46 mmHg ($40 + 0.7 \times 8$) (15). The equation for this is as follows:

$$Compensated\ pCO_2 = Initial\ pCO_2 + 0.7$$

$$\times \left[Compensated\ HCO_3^- - Initial\ HCO_3^- \right] \quad (2.3)$$

$$pCO_2 = 40 + 0.7 \times \left[HCO_3^- - 24 \right] = 40 + 0.7 \times 8 = 46$$

If the actual pCO_2 deviates from this expected pCO_2 by more than 4 mmHg, the patient may have a mixed disorder due to either an underlying respiratory acidosis (if more than 4 mmHg higher) or alkalosis (if more than 4 mmHg lower).

Caution: These calculations for expected compensation should be used as guidelines in the diagnosis and should not be overinterpreted as an absolute diagnosis.

Treatment. The optimal treatment for metabolic alkalosis is correcting the underlying disorder and any factors that perpetuate the disorder. In some cases, simple hydration will gradually correct metabolic alkalosis if renal function is normal. Because Cl^- depletion is commonly present, Cl^- may be replaced by the administration of NaCl or KCl solutions as appropriate, with consideration of the renal function. Rarely, dilute HCl may be administered intravenously if necessary, and Brandis points out that correction of metabolic alkalosis will cause a right shift in the oxygen dissociation curve, which will increase peripheral oxygen release (16,17). In patients with anasarca or congestive heart failure who remain in an edematous state and continue to require diuresis, acetazolamide, a carbonic anhydrase inhibitor, may be given. Acetazolamide primarily inhibits proximal sodium bicarbonate reabsorption in the kidney, thereby increasing urinary bicarbonate excretion.

Respiratory acidosis

Arterial pCO_2 is a balance between cellular production and removal by alveolar ventilation. Respiratory acidosis is usually caused by inadequate ventilation that results in excess CO_2 in the blood and a concurrent decrease in blood pH below 7.35. Acute respiratory acidosis is most commonly related to decreased ventilation, but can also be caused by increased production of CO_2 by the body, or excess CO_2 in the inspired gas [18]. Normally, respiratory centers in the pons and medulla control alveolar ventilation by chemoreceptors sensitive to changes in pCO_2, pO_2, and pH, such that any increased production of CO_2 promptly stimulates hyperventilation, which maintains arterial pCO_2 at normal levels [19]. If arterial pCO_2 is increased, it usually involves a problem with ventilation.

Mechanisms of acute and chronic respiratory acidosis:

- Ventilation failure due to CNS depression caused by brain injury, drugs such as opioids or anesthetics, or respiratory muscle insufficiency as seen in neuromuscular disorders, such as myasthenia gravis, amyotrophic lateral sclerosis, or muscular dystrophy.
- Impaired pulmonary gas exchange due to a high dead space fraction [(total ventilation—alveolar ventilation)/total ventilation]. This can be seen in a number of pulmonary diseases associated with the destruction of the alveolar space or impaired blood flow to a segment of the lung. In these scenarios, effective alveolar ventilation makes up a smaller fraction of the total minute ventilation.
- Chronically decreased reflex responsiveness to hypoxia and hypercapnia seen in COPD, chronic restrictive lung disease, and obesity hypoventilation syndrome [19].
- Acute airway obstruction caused by aspiration of a foreign object or blockage of an endotracheal tube
- Circulatory failure, which causes insufficient delivery of blood to the lungs
- Increased metabolism or CO_2 production in a mechanically ventilated patient who is not spontaneously breathing. If CO_2 production increases while ventilation remains constant, the blood pCO_2 will rise [20].

Expected compensation in respiratory acidosis [8,20] (Table 2.3 and Fig. 2.1). To compensate the excess CO_2 and acid in the blood, the kidneys begin to excrete more hydrogen and ammonium ions and reabsorb more bicarbonate. During the acute phase, plasma buffering of the elevated CO_2 increases the HCO_3^- slightly, by approximately 1 mmol/L for each 10 mmHg rise in pCO_2. Over the next several hours, the kidneys increase the reabsorption of HCO_3^-, which elevates serum HCO_3^- by about 2 mmol/L. The kidneys can also increase bicarbonate retention by urinary excretion of a high chloride (and ammonium ion) concentration relative to sodium, which would lessen bicarbonate loss. As this continues into the chronic phase (over 24 h), renal compensation occurs gradually over several days to increase HCO_3^-, reaching a plateau after 2–5 days. Even after several days, compensation generally does not return pH completely back to normal *(8)*. As compensation occurs in two phases, the expected compensation has two algorithms to predict the expected blood level of HCO_3^- *(21)*.

Acute Resp Acid: For each 10 mmHg rise in arterial pCO_2 above 40 mmHg, HCO_3^- should increase by about 1 mmol/L and pH should decrease by about 0.07–0.08. For example, if a patient's breathing slows for a few hours and the pCO_2 increases from 40 to 50 mmHg, the HCO_3^- should rise by about 1 mmol/L (i.e., 25 mmol/L) and the pH should decrease by 0.08. For example, pH decreases from 7.40 to about 7.32.

$$\text{Compensated } HCO_3^- = \text{Initial } HCO_3^- + 0.1 \times (\text{Compensated } pCO_2 - \text{Initial } pCO_2) \quad (2.4)$$

$$\text{pH change} = 0.008 \times (40 - pCO_2) = 0.008 \times -10 = -0.08 \quad (2.5)$$

Chronic Resp Acid: The maximal HCO_3^- response occurs in about 3 days, with the HCO_3^- increasing by 3–4 mmol/L for each 10 mmHg rise in arterial pCO_2. For example, if the pCO_2 has increased from a patient's baseline of 40 mmHg to a steady state of 60 mmHg for several days, the HCO_3^- should have risen about 6–8 mmol/L. In our example, if the patient's normal HCO_3^- was 24 mmol/L, it should be about 30–32 mmol/L after 3 days of metabolic compensation. Thus, for a pCO_2 increase from 40 to 60 mmHg, the expected bicarbonate during compensation may be calculated by

$$\text{Compensated } HCO_3^- = \text{Initial } HCO_3^- + 0.35 \times (\text{Compensated } pCO_2 - \text{Initial } pCO_2) \quad (2.6)$$

or:

$$HCO_3^- = 24 + 0.35 \times (60 - 40) = 31 \text{ mmol/L}$$

The expected pH change may be calculated by

$$\text{pH change} = 0.003 \times (40 - pCO_2) = -0.06 \quad (2.7)$$

Note that these calculations are time dependent. If known, the patient's baseline or initial values are preferred, but if not known, we may assume the patient started with a very normal pH (7.40), pCO_2 (40 mmHg), and HCO_3^- (24 mmol/L).

Caution: These calculations for expected compensation should be used as guidelines in the diagnosis and should not be overinterpreted as an absolute diagnosis.

Treatment of respiratory acidosis. The increased blood pCO_2 should be corrected gradually to prevent sudden alkalinization of the CSF that could lead to seizures. While ventilation with positive pressure ventilation may be necessary, less-invasive pharmacologic therapy targeting the underlying cause of respiratory acidosis such as bronchodilators or opioid blockers such as naloxone may be adequate (19). Administration of bicarbonate is rarely done, because HCO_3^- crosses the blood−brain barrier slowly, which can elevate the blood pH without affecting the CNS pH (22).

Respiratory alkalosis

Perhaps surprisingly, respiratory alkalosis has been cited as the most common acid−base disorder (12). Respiratory alkalosis can be caused by any condition that leads to hyperventilation and is associated with a fall in blood pCO_2 to below 35 mmHg and an elevated blood pH. From a normal minute ventilation of 5−8 L/min, respiratory alkalosis (hyperventilation) results from an increased minute ventilation, in which CO_2 is lost faster than it is produced, leading to an increased pH and decreased pCO_2. Normally, respiratory centers in the pons and medulla control ventilation, with many factors inducing hyperventilation.

Respiratory alkalosis may be an acute or chronic process. While hyperventilation can change pCO_2 levels very quickly, metabolic compensation by increasing loss of HCO_3^- and increasing retention of H ions is a much slower process and takes several hours to days to compensate the pH imbalance. Therefore, HCO_3^- levels are mostly normal in acute hyperventilation because renal excretion has not had sufficient time to lower the HCO_3^- level. As hyperventilation enters the chronic phase, HCO_3^- levels will be decreased.

Mechanisms of respiratory alkalosis (12,23):

- CNS-induced hyperventilation by head trauma, stroke, pain, anxiety, sepsis, fear, stress, drugs and medications.
- Hypoxemia-induced hyperventilation, as caused by pulmonary embolism, pneumonia, pneumothorax, and acute exacerbations of chronic lung diseases.
- Excess minute ventilation with a positive pressure ventilator.

Expected compensation in respiratory alkalosis (9,23,24) (Table 2.3 and Fig. 2.1). The expected metabolic response to the decreased blood pCO_2 in respiratory alkalosis is the kidney decreases reabsorption of bicarbonate in the proximal tubule, which lowers plasma HCO_3^- and decreases plasma pH. As metabolic compensation is relatively slow, it is time dependent and occurs in two phases. Thus, the expected compensation has two algorithms to predict the expected blood level of HCO_3^-:

Acute Resp Alkalosis (<24 h): For each 10 mmHg fall in arterial pCO_2 below 40 mmHg, the pH should increase by about 0.08, and HCO_3^- should decrease by ∼2 mmol/L. As an example, during this phase, if pCO_2 decreases from 40 to 30 mmHg, this 10 mmHg fall should be associated with a pH increase of 0.08 (pH ∼ 7.48) and a bicarbonate decrease of 2 mmol/L, from 24 to ∼22 mmol/L. The equation for this is:

$$\text{Compensated } HCO_3^- = \text{Initial } HCO_3^- -0.2 \\ \times (\text{Initial } pCO_2 - \text{Compensated } pCO_2) \quad (2.8)$$

Chronic Resp Alkalosis (2−5 days): For each 10 mmHg fall in pCO_2, pH should increase by about 0.03 and HCO_3^- should decrease by about 5 mmol/L. For example, if the pCO_2 decreases from 40 to 30 mmHg after 2 days of hyperventilation, the pH should increase by 0.03 (pH ∼7.43) and HCO_3 should decrease by about 5, from 24 to 19 mmol/L.

$$\text{Compensated } HCO_3^- = \text{Initial } HCO_3^- -0.5 \\ \times (\text{Initial } pCO_2 - \text{Compensated } pCO_2) \quad (2.9)$$

- **Note:** The lower limit for metabolic compensation in hyperventilation is a plasma HCO_3^- of ∼12 mmol/L (∼12 mEq/L). If plasma HCO_3^- is <12 mmol/L (<12 mEq/L), an underlying metabolic acidosis may be present.

Caution: These calculations for expected compensation should be used as guidelines in the diagnosis and should not be overinterpreted as an absolute diagnosis.

Treatment of respiratory alkalosis. Treatment is aimed at correcting or alleviating the underlying condition, such as medications for anxiety or pain, antibiotics for sepsis, or anticoagulants for a pulmonary embolism *(12)*. Hypoxemia must be corrected by providing supplemental oxygen. Anxiety-induced hyperventilation may be treated with reassurance or sometimes by rebreathing expired air (higher CO_2 content) in otherwise healthy persons *(23)*.

Detecting mixed acid−base disorders

Mixed acid−base disorders occur when multiple primary acid−base disorders occur at the same time. These are common in hospital populations, especially in the emergency department and the intensive care unit. Although the equations and Figs. 2.1 and 2.2 can aid in the diagnosis, these relationships may not hold perfectly in patients with complex and chronic acid−base disorders that change during the disease process. There is no substitute for a careful review of a patient's history and clinical evaluation.

The importance of identifying mixed acid−base disorders lies in their diagnostic and therapeutic implications. For example, the development of a

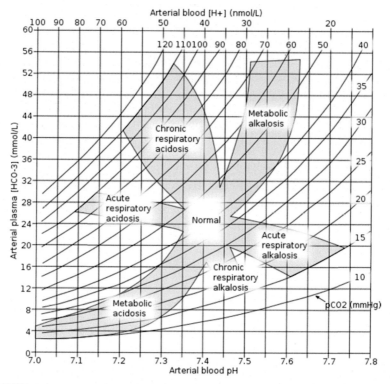

FIGURE 2.2 Nomogram for interpreting acid–base status. Y-axis is plasma bicarbonate in mmol/L. The lower X-axis is pH; the upper X-axis is H ion concentration in nmol/L. The curved lines represent constant pCO_2 values in mmHg. Curved lines represent a constant pCO_2. *Met Alk* metabolic alkalosis; *Resp Acid* respiratory acidosis; *Resp Alk* respiratory alkalosis. *Used with public permission from Huckfinne, Public domain, via Wikimedia Commons. https://commons. wikimedia.org/wiki/File:Acid-base_nomogram.svg.*

primary metabolic alkalosis in a patient with chronic obstructive airway disease who is being treated with diuretics should alert the clinician to possible potassium depletion, and a patient presenting with a mixed respiratory alkalosis and metabolic acidosis should be evaluated for possible salicylate intoxication *(9,25,26).*

Does the expected compensation occur?

A simple concept to remember in a primary acid–base disorder is that, if the expected compensation does not occur, a mixed disorder should be suspected. For example, in a primary metabolic acidosis, the lungs are expected to compensate this excess metabolic acid by hyperventilating (a respiratory alkalotic process) to remove the respiratory acid CO_2 and return the pH toward normal. If this does not occur as expected, the person is considered to have an underlying respiratory acidosis *(9,24−26)*.

Metabolic acidosis: The respiratory response to compensate metabolic acidosis is hyperventilation, which normalizes the pH by lowering the pCO_2 to balance the deficit of HCO_3^-. This compensating hyperventilation starts within 30 min and should be maximal by 12−24 h. This response by hyperventilation should decrease the blood pCO_2 by 1.2 mmHg for each 1 mmol/L drop in blood HCO_3^-. As derived earlier *(10)*, the expected pCO_2 (mmHg) during this time should be within 3 mmHg of

$$\text{Compensated } pCO_2 = \text{Initial } pCO_2 - 1.2 \times \left[\text{Initial } HCO_3^- - \text{Compensated } HCO_3^-\right] \quad (2.10)$$

Metabolic alkalosis: The compensating response to metabolic alkalosis is hypoventilation, which increases the pCO_2. As described earlier *(15)*, during this period of compensation, the expected pCO_2 (mmHg) should be within 3 mmHg of

$$\text{Compensated } pCO_2 = \text{Initial } pCO_2 + 0.7 \times \left[\text{Compensated } HCO_3^- - \text{Initial } HCO_3^-\right] \quad (2.11)$$

Fig. 2.1 shows the expected areas of pCO_2 versus HCO_3^- for both metabolic acidosis and metabolic alkalosis. However, as mentioned previously, expected compensation calculated by an equation should only be used as a guide and not as an absolute diagnosis. This is especially true if a patient has had multiple (mixed) acid−base disorders occurring over a period of time.

Respiratory acidosis or respiratory alkalosis: If a patient has a simple respiratory acidosis, the expected compensated HCO_3^- concentration may be predicted by Eq. (2.4) or (2.5) for acute respiratory acidosis and by Eq. (2.6) or (2.7) for chronic respiratory acidosis. If a patient has a simple respiratory alkalosis, the expected compensated HCO_3^- concentration may be predicted by Eq. (2.8) for acute changes and by Eq. (2.9) for chronic changes. Also, simple rules predict the expected pH changes versus the change in pCO_2 during the acute and chronic phases. During the acute phase of respiratory disorders, for each 10 mmHg rise or fall in pCO_2, the pH should change by 0.08. As this progresses into the chronic phase (1−2 days), for each 10 mmHg rise or fall in $pCO2$, the pH should have changed by only 0.03 *(24,25)*. Table 2.4 illustrates the relationships between pCO_2, pH, and HCO_3^- for acute and chronic respiratory changes:

Fig. 2.1 is a graphical plot of the equations for predicting how bicarbonate is expected to change during acute and chronic primary respiratory disorders and how pCO_2 is expected to change during primary metabolic disorders with compensation. If the blood gas data clearly do not reflect these temporal relationships, it suggests an additional disorder is present and that the patient has

TABLE 2.4 Expected changes for pCO_2, pH, and HCO_3^- for acute and chronic respiratory conditions.

pCO_2 (mmHg)	Expected changes in acute respiratory conditions		Expected changes in chronic respiratory conditions	
	pH	HCO_3^- (mmol/L)	pH	HCO_3^- (mmol/L)
70	7.16	27	7.31	34.5
60	7.24	26	7.34	31
50	7.32	25	7.37	28
40	**7.40**	**24**	**7.40**	**24**
30	7.48	22	7.43	19
20	7.56	20	7.46	14

a mixed disorder. However, these relationships are guidelines only and may not hold for complex acid–base disorders.

Delta gap and delta ratio *(5,9,27–29)*. The "delta gap," is the difference between the increase in AG and the decrease in HCO_3^-. It can help determine if the high anion gap metabolic acidosis is solely explained by the decrease in HCO_3^-, or if a possible mixed acid–base disorder is present. If the delta gap is close to zero, it is likely that a single or "pure" metabolic acidosis is present. If the delta gap is clearly positive, it suggests that there is more HCO_3^- in the blood (i.e., less HCO_3^- is lost) than expected from the increased AG so that an additional metabolic alkalosis may be present. If the delta gap is clearly negative, it suggests less HCO_3^- is present (more HCO_3^- is lost) than expected from the increased AG, so that an additional non-AG metabolic acidosis may be present.

$$\text{Delta Gap} = (AG - 12) - (24 - HCO_3^-) \qquad (2.12)$$

The "delta ratio" is based on the same concept for interpreting an elevated AG, except that it uses the ratio of the increase in AG ($Na-Cl-HCO_3^-$) from a midnormal value of 12, to the decrease in HCO_3^- from a normal value of 24. An equation for delta ratio is as follows:

$$\text{Delta Ratio} = (AG - 12) / (24 - HCO_3^-) \qquad (2.13)$$

[Note that a HCO_3^- value of 24 makes the delta ratio undefined because the denominator is zero] In a typical metabolic acidosis, the increase in AG from 12 mmol/L and the decrease in HCO_3^- from 24 mmol/L should approximately equal each other, so the "delta ratio" would be approximately 1. If the delta ratio is appreciably greater than 1, it suggests that there is more HCO_3^- in the blood than expected from the increased AG, so that an additional metabolic alkalosis may be present. If the delta ratio is appreciably less than 1, it means less HCO_3^- is present than expected from the increased AG, so this suggests an additional non-AG metabolic acidosis is present. Examples of various delta ratios and delta gaps are illustrated in Table 2.5, and interpretive ranges are shown in Table 2.6.

Here are some findings that strongly suggest mixed acid–base disorders:

- If the pH is well within normal limits and both the HCO_3^- and pCO_2 are abnormal.
- HCO_3^- and pCO_2 are abnormal in the opposite direction, as for example, an elevated HCO_3^- with a decreased pCO_2. The compensatory response should always be in the same direction as the change caused by the primary disorder.
- Simple acid–base disorders do not overcompensate. That is, compensation will not cause an acidemia to become an alkalemia. In fact, the compensation mechanism typically stops slightly before the pH normalizes (9).

TABLE 2.5 Examples of delta ratios and delta gaps in metabolic acidoses.

Anion gap (mmol/L)	HCO$_3$ (mmol/L)	Delta ratio	Delta gap	Condition(s)
12	24	(undefined)	0	Normal values
18	24	(undefined)	+8	Mixed disorder: AG metabolic acidosis + metabolic alkalosis
24	18	2.0	+6	Mixed disorder: AG metabolic acidosis + metabolic alkalosis
20	16	1.0	0	Pure AG metabolic acidosis
24	12	1.0	0	Pure AG metabolic acidosis
18	13	0.55	−5	Mixed disorder: AG metabolic acidosis + non-AG metabolic acidosis
20	10	0.57	−6	Mixed disorder: AG metabolic acidosis + non-AG metabolic acidosis

Delta ratio = (AG − 12)/(24 − HCO$_3$)
Delta gap = (AG − 12) − (24 − HCO$_3$)
Delta gap = AG + HCO$_3$ − 36
Delta gap = Na − Cl − 36

TABLE 2.6 Interpretation of delta gap and delta ratios when metabolic acidosis is present.

Delta gap	Delta ratio	Interpretation
−5 or less	0.5–1.9	Mixed anion gap acidoses: AG metabolic acidosis + non-AG metabolic acidosis
−4 to +4	0.4	A "pure" high anion gap acidosis
+5 or greater	2.0 or greater	A mixed high AG acidosis + metabolic alkalosis

Diagnostic approach to acid–base disorders

Reference intervals

The reference intervals for both arterial and venous blood are shown in Table 1.1 for the common blood gas and acid–base parameters (30,31).

Stepwise approach to evaluating acid–base status

Ideally, evaluating acid–base disorders, including mixed acid–base disorders, would include simultaneous information on the patient's clinical history, physical exam, electrolyte results, and blood gas results. With modern blood gas analyzers, the blood gas results are usually available with electrolyte, metabolite, and hemoglobin results.

Fig. 2.2 is helpful as a guide to evaluate the patient's pH, pCO_2, and HCO_3 results for a potential acid–base disorder. The panel of blood gas results (pH, pCO_2, pO_2, HCO_3, Na, K, lactate et al.) may be evaluated in several steps, as follows:

(1) Evaluate the patient's clinical history and current status to anticipate conditions associated with acid–base disorders (see Table 2.7).
(2) Evaluate the pH.
(3) Evaluate the ventilatory (pCO_2) and metabolic (HCO_3^-) status to gain insight into whether the acid–base disorder is simple or possibly a mixed disorder.
(4) a. Evaluate laboratory and clinical data for a possible mixed disorder and if compensation is appropriate.
   b. Do parameters such as electrolytes, lactate, AG, or delta gap/ratio indicate that another acid–base disturbance is present?
   c. Is the patient's clinical presentation consistent with the blood gas results?

The evaluation of oxygen status is mostly independent of the acid–base interpretation, and it is discussed in Chapter 4.

Step 1. Evaluate the patient's clinical history and status to anticipate possible acid–base disorders

There are many clinical conditions associated with acid–base disorders, as listed in Table 2.7. A perceptive clinical evaluation, both at initial evaluation and during the clinical course, is an invaluable part of evaluating a patient for acid–base disorders.

TABLE 2.7 Examples of the expected acid–base disorder associated with various clinical conditions.

Clinical condition	Expected acid–base disorder
Cardiac arrest	Metabolic acidosis
Pulmonary arrest	Respiratory acidosis
Hyperventilation (many causes)	Respiratory alkalosis
Blockage of endotracheal tube	Respiratory acidosis
Vomiting	Metabolic alkalosis
Diarrhea	Metabolic acidosis
Shock state (inadequate perfusion)	Metabolic acidosis
Acute pulmonary edema	Respiratory alkalosis (hypoxia leads to hyperventilation)
Severe pulmonary edema	Respiratory acidosis
Diuretic therapy	Metabolic alkalosis
Drug intoxication	Respiratory acidosis (respiratory arrest)
Bicarbonate therapy	Metabolic alkalosis
Poor perfusion	Metabolic acidosis

Step 2. Evaluate the pH

An abnormal pH indicates that an acidosis or alkalosis has occurred and the extent of the acid–base disorder, and it may suggest that compensation has occurred. However, the pH by itself does not indicate whether a mixed disorder is present. Consider the following examples:

- pH 7.20 confirms that a severe acidosis is present and that compensation is either in its early stages or is ineffective in controlling the acidosis. Further investigation is required to determine the metabolic or respiratory origin of the acidosis and whether more than one acidotic process is present.
- pH 7.48 indicates a mild alkalosis. Further information is required to determine the cause of the alkalosis and whether the alkalosis is in its early

stages (and may get worse), has been nearly compensated, or is part of a mixed disorder.

A normal pH may indicate that the patient has no acid–base disorder. However, the patient may have a mix of acidotic and alkalotic events (primary or compensatory) that have offset each other. In addition to the patient's history, the HCO_3^- and pCO_2 must be considered, as in these examples:

- pH 7.45 with elevated HCO_3^- and pCO_2 suggests that a primary metabolic alkalosis has been compensated by a respiratory acidotic response. Similarly, a pH of 7.45 with decreased HCO_3^- and pCO_2 suggests that a primary respiratory alkalosis has been compensated by a metabolic acidotic response.
- pH 7.40 with both HCO_3^- and pCO_2 abnormal indicates that a mixed acidosis and alkalosis are present that have fortuitously offset each other to give a normal pH.

Step 3. Evaluate the ventilatory and metabolic status

The simplest approach to evaluating the pCO_2 and HCO_3^- is to consider each parameter separately as indicating acidosis, alkalosis, or normal status, then evaluate them along with pH to determine whether the primary disorder is respiratory or metabolic and whether a mixed disorder or compensation is present.

A decreased pCO_2 indicates a respiratory alkalotic process, which may be either primary or compensatory. An increased pCO_2 indicates a respiratory acidotic process, either primary or compensatory (Table 2.8).

A decreased HCO_3^- indicates a metabolic acidosis, either primary or compensatory. An increased HCO_3^- indicates a metabolic alkalosis, either primary or compensatory (Table 2.9).

Note: The terms "respiratory alkalosis," "metabolic acidosis," etc., refer to pathologic processes and not to compensation. However, I believe that it simplifies interpretation to momentarily regard any decreased pCO_2 (e.g.) as a respiratory alkalotic process. Whether it is a primary abnormality or appropriate compensation will become apparent as other factors are considered.

TABLE 2.8 Evaluation of ventilatory status by arterial pCO_2.

pCO_2	Ventilatory status
Decreased	Respiratory alkalotic process (hyperventilation) is present
Normal	Normal ventilatory status? (Is a mixed disorder present?)
Increased	Respiratory acidotic process (hypoventilation) is present

TABLE 2.9 Evaluation of metabolic status by plasma bicarbonate (HCO_3^-).

Bicarbonate (total CO_2)	Metabolic status
Decreased	A metabolic acidotic process is present (either primary, compensatory, or mixed)
Normal	No metabolic acid–base disorder present? (mixed disorder also possible)
Increased	A metabolic alkalotic process is present (either primary, compensatory, or mixed)

If either pCO_2 or HCO_3^- remain normal when the other has been abnormal for a sufficient period of time, then a mixed disorder may be present. That is, a lack of an appropriate compensatory response suggests an additional disorder is present.

Example 1. A pCO_2 result of 25 mmHg indicates that a respiratory alkalotic process is present. Whether it is a primary abnormality or appropriate compensation depends on the HCO_3^- and pH associated with this pCO_2 and the timing of the results. If the HCO_3^- is 14 mmol/L (14 mEq/L) with a slightly acidic pH of ∼7.32 for example, then the pCO_2 of 25 mmHg probably indicates compensating hyperventilation (a respiratory alkalotic process) for a primary metabolic acidosis.

Example 2. For a similar pCO_2 of 25 mmHg, if the HCO_3^- is a normal 23 mmol/L (23 mEq/L) and the pH is alkaline at ∼7.53, the decreased pCO_2 suggests an acute primary respiratory alkalosis. If the HCO_3^- is 16 mmol/L (16 mEq/L), that suggests a chronic respiratory alkalosis. The duration of the hyperventilation should also be evaluated in determining whether it is acute or chronic hyperventilation. If acute, the HCO_3^- of 23 mmol/L suggests the metabolic compensatory response to eliminate HCO_3^- is progressing slowly as expected. If the hyperventilation is chronic, then the HCO_3^- of 23 mmol/L is higher than expected, suggesting that, because the kidney is unable to adequately compensate by eliminating HCO_3^-, a metabolic alkalosis may also be present.

Step 4. Evaluate laboratory and clinical data for a possible mixed disorder

Step 4 has three components.

Step 4a. Is the compensation appropriate for the primary disorder?

To evaluate if the expected compensation is occurring for the primary acid–base disorder, refer to the "Expected Compensation" descriptions in the various sections of acid–base disorders described in this chapter and the section on "Detecting Mixed Acid–Base Disorders."

Fig. 2.1 is a nomogram derived from equations listed here and reported elsewhere *(1,32)* that should aid in understanding and interpreting the changes in acid–base results that occur in the various disorders. While both Figs. 2.1 and 2.2 are useful, the actual diagnosis of a patient's condition must always include a clinical assessment of the patient's history and physical exam findings at the time the result was obtained. Figs. 2.1 and 2.2 are helpful in simple or mixed double disorders that occur simultaneously, but these figures become less reliable when one or more new disorders develop in addition to an existing disorder.

Step 4b. Do other laboratory results suggest an additional acid–base disorder is present?

Several common laboratory tests may help determine whether additional acid–base disorders are present *(25,26,33)*.

Potassium. Hypokalemia suggests a metabolic alkalosis may be present.

pH. A normal pH combined with abnormal results for both HCO_3^- and pCO_2 warrants consideration of a mixed acid–base disorder.

Chloride. Hyperchloremia may result from several causes as described under metabolic acidosis, such as renal tubular acidosis in which HCO_3^- is lost in the urine, either from a primary metabolic acidosis or from the metabolic acidotic compensation for respiratory alkalosis. Hypochloremia promotes renal retention of HCO_3^-, leading to metabolic alkalosis. Compensation for chronic respiratory acidosis increases renal HCO_3^- retention with loss of Cl^-.

Anion gap (AG), calculated from measurements of Na, (K), Cl, and HCO₃. An elevated AG [(Na)−(Cl + HCO₃)] of >20 mmol/L (>20 mEq/L) indicates metabolic acidosis. AG is also useful in differentiating the causes of metabolic acidoses, as shown in Table 2.1 *(5,34)*.

- Elevated AG occurs in uremic acidosis, ketoacidosis, lactic acidosis, and ingestion of certain medications or toxins (e.g., ethylene glycol or salicylate).
- Normal AG occurs in renal tubular acidosis and diarrhea, among other conditions.

While often very useful, AG is calculated from three or four measurements and is subject to variation from these measurements. Furthermore, a correction for abnormal albumin concentrations is now recommended, as noted in Chapter 1 and Table 2.2. Several reports have found that about half of patients with a clearly elevated blood lactate did not have an elevated AG *(5,35)*.

Delta gap or delta ratio. The delta gap and delta ratio are discussed earlier in the section "Detecting Mixed Acid–Base Disorders." If the delta gap is clearly positive or if the delta ratio is approximately two or greater, it suggests there is more HCO_3^- in the blood (i.e., less HCO_3^- is lost) than expected from the increased AG, so that an additional metabolic alkalosis may be present. If the delta gap is clearly negative or if the delta ratio is approximately 0.4–0.8, it suggests less HCO_3^- is present (more HCO_3^- is lost) than expected from the increased AG, so that an additional non-AG metabolic acidosis may be present *(28,29)*.

Lactate. A rising blood lactate is a sensitive indicator of metabolic acidosis resulting from a variety of oxygenation disorders related to ventilation, alveolar gas exchange, circulatory transport, cellular uptake, and/or mitochondrial dysfunction.

Creatinine. An elevated creatinine indicates renal insufficiency and possible uremic metabolic acidosis.

Step 4c. Does the patient have other conditions associated with an acid–base disorder?

As noted in Step 1, an evaluation of acid–base status must include the patient's historical and physical exam findings. Many conditions are frequently associated with acid–base disorders, with some common ones listed in Table 2.7. Especially if the blood gas results do not readily explain the patient's "primary" disorder, this should alert the clinician to the possibility of a mixed acid–base disorder. Mixed acid–base conditions are quite common, with some patients having two or three independent acid–base disorders *(9)*.

Summary of acid–base (pH, pCO_2, and HCO_3^-) interpretations

- Evaluate the pH: Above 7.45 = Alkalosis Below 7.35 = Acidosis
- Evaluate ventilatory (pCO_2) and metabolic (HCO_3^-) status:
    pH and pCO_2 change in same direction = Metabolic disorder
    pH and pCO_2 change in opposite direction = Respiratory disorder
- Determine if a mixed disorder is present:
    Is renal or pulmonary compensation occurring when expected?
        Yes: probably a simple disorder No: probably a mixed disorder
    Do HCO_3^- and pCO_2 change in opposite directions? Yes: probably a mixed disorder
    Do other tests suggest another acid–base disorder is present?
        K, Cl, AG, Lactate, Creatinine, delta gap/ratio.
- Does the patient's history indicate another acid–base disorder is present?

Complex acid—base case example *(9)*

Okay class, since you should all be experts in blood gas knowledge by now, a good example of how a complex acid—base case might be evaluated is provided by Emmett et al. *(9)*. It demonstrates several concepts: (1) the challenge of evaluating the compensatory response in respiratory acid—base disorders because they are time dependent; (2) the various nomograms that may be useful as a guide, but must be interpreted with caution in complex acid—base disorders; and (3) the importance of evaluating the patient's history. Here are the acid—base results:

$$\text{pH: } 7.27 \quad p\text{CO}_2\text{: } 70 \text{ mmHg } (9.3 \text{ kPa}) \quad \text{HCO}_3^-\text{: } 31 \text{ mmol/L}$$

The low pH and increased $p\text{CO}_2$ indicate the person has a respiratory acidosis. According to Fig. 2.2, for a $p\text{CO}_2$ of 70 mmHg, the compensated HCO_3^- should be $\sim$27 mmol/L for acute respiratory acidosis or about 36 mmol/L for chronic respiratory acidosis. Because the measured HCO_3^- is between these two concentrations, a mixed acid—base disorder is likely. Here are four possible explanations:

- Chronic respiratory acidosis + metabolic acidosis. This could be a patient with chronic obstructive pulmonary disease who developed a metabolic acidosis from either diarrhea or impaired oxygen metabolism in sepsis. The correct diagnosis could be determined if the patient history confirmed recent diarrhea, or if their lactate concentration was elevated due to sepsis.
- A patient with chronic respiratory acidosis who developed a superimposed acute respiratory acidosis. The patient might have had a $p\text{CO}_2$ of 55 mmHg (7.3 kPa) with an appropriately compensated HCO_3^- of 31 mmol/L, then developed pneumonia that acutely increased the $p\text{CO}_2$ to 70 mmHg.
- Acute respiratory acidosis + metabolic alkalosis. This could be a person who developed acute respiratory depression from taking a sedative, then developed a metabolic alkalosis from either vomiting or taking a diuretic that increased the HCO_3^- from 27 to 31 mmol/L.
- Acute respiratory acidosis that is evolving into a chronic respiratory acidosis. This could occur about 1—3 days after the acute episode began.

Self-assessment and mastery

Acid—base exercises: In the following table, use the pH, $p\text{CO}_2$, and HCO_3^- values along with the duration of observation to assess the acid—base status of each situation. Determine the primary disorder, then evaluate if the expected compensation has occurred for the duration indicated. Use Figs. 2.1 and 2.2 for guidance as needed:

pH	pCO_2 (mmHg)	HCO_3 (mmol/L)	Duration of medical observation	Acid–base condition
7.40	40	24	–	Very normal acid–base results
7.34	34	18	12 h	
7.55	49	40	6 h	
7.55	49	40	24 h	
7.44	27	18	2 days	
7.23	70	27	6 h	
7.23	70	29	2 days	
7.28	55	25	2 h	
7.41	70	43	2 h	
7.52	31	29	6 h	
7.08	54	18	12 h	
7.56	20	20	6 h	

Answers to Acid–Base Exercises:

pH	pCO_2 (mmHg)	HCO_3 (mmol/L)	Duration of medical observation	Acid-base condition
7.40	40	24	–	Very normal acid–base results
7.34	34	18	12 h	Partially compensated metabolic acidosis
7.55	49	40	6 h	Metabolic alkalosis
7.55	49	40	24 h	Combined metabolic alkalosis and respiratory alkalosis
7.44	27	18	2 d	Chronic (compensated) respiratory alkalosis
7.23	70	27	6 h	Acute respiratory acidosis
7.23	70	29	2 d	Primary respiratory acidosis with a metabolic acidosis preventing an increase in HCO_3
7.28	55	25	2 h	Acute respiratory acidosis
7.41	70	43	2 h	Combined respiratory acidosis and metabolic alkalosis
7.52	31	29	6 h	Combined (mixed) metabolic and respiratory alkalosis
7.08	54	18	12 h	Combined (mixed) metabolic and respiratory acidosis
7.56	20	20	6 h	Acute respiratory alkalosis

Delta gap/ratio exercises: In the following table of possible results for metabolic acidoses, use the Anion Gap (AG) and HCO_3 values to calculate the delta gaps and delta ratios. Assume a normal AG is 12 mmol/L and a normal HCO_3 is 24 mmol/L.

AG	HCO$_3$	Delta gap	Delta ratio
14	18		
16	18		
18	18		
24	18		
24	14		
24	20		

Answers to delta gap/ratio exercises:

AG	HCO$_3$	Delta gap	Delta ratio
14	18	−4	0.33
16	18	−2	0.66
18	18	0	1.0
24	18	+6	2.0
24	14	+2	1.2
24	20	+8	3.0

Case examples

Assess the following patients for their acid—base, ventilatory, and oxygenation status. Remember to assess the clinical picture for clues and determine the anion gap and delta ratio where appropriate.

Case 1. A 54-y-old woman with a history of hypertension and aortic valve stenosis developed severe lower extremity edema and shortness of breath. She was admitted for aortic valve replacement. After this operation she remained on a ventilator for several days, but is now extubated and breathing 50% oxygen. Her ventilatory mechanics are marginal to poor. Her arterial blood gas results are as follows:

pH	7.28
pCO$_2$	53 mmHg
pO$_2$	116 mmHg
HCO$_3^-$	25 mmol/L (25 mEq/L).
FI-O$_2$ (%)	50%

Case 2. A female infant was operated on at 7 d of age to correct pulmonary atresia and a patent ductus arteriosus. A central shunt was placed from the right ventricle to the pulmonary artery. Several days after the operation, she developed congestive heart failure and pulmonary edema due to excess shunt flow. Solution of NaHCO$_3$ was administered to correct low pH caused by lactic

acidosis due to inadequate peripheral perfusion. At 11 d of age, the central
shunt was revised with a smaller diameter shunt. At 10 and 15 d of age she was
breathing 40% oxygen when the following arterial blood gas results were
obtained:

	Day 10	Day 15
pH	7.48	7.44
pCO_2	46 mmHg	44 mmHg
pO_2	40 mmHg	60 mmHg
HCO_3^-	34 mmol/L (33 mEq/L)	30 mmol/L (30 mEq/L)
Hb	102 g/L (10.2 g/dL)	108 g/L (10.8 g/dL)
Lactate	5.8 mmol/L (52 mg/dL)	1.5 mmol/L (13.5 mg/dL)

Case 3. A young boy with a history of asthma is admitted to the ED with a
2-day history of difficulty breathing and a respiratory rate of 28/min (normal
about 13/min). He is alert but is wheezing and has great difficulty speaking in
more than one to two word sentences. His lab results are:

pH	7.40
pCO_2	28 mmHg
pO_2	45 mmHg
sO_2	70%
Na:	134 mmol/L
K	4.9 mmol/L
Cl	96 mmol/L
HCO_3^-	16 mmol/L (16 mEq/L)

Case 4. A 32-year-old man with diabetes and renal failure was admitted for
renal transplantation. Intraoperative blood gas results were as follows while
breathing 60% oxygen:

pH	7.20
pCO_2	43 mmHg
pO_2	178 mmHg
HCO_3^-	17 mmol/L (17 mEq/L)

After the operation, the patient was successfully extubated. The following
results were obtained 24 h after extubation while he was breathing room air:

pH	7.33
pCO_2	36 mmHg
pO_2	82 mmHg
HCO_3^-	19 mmol/L (19 mEq/L)

Case 5. About 1 month after receiving a liver transplant, a 56-year-old man had developed respiratory distress with hyperventilation lasting over a week. On admission, the patient had a 1.5-L pleural effusion that was drained. He was stabilized with an FiO_2 of 0.40 while the following blood gas results were obtained:

pH	7.40
pCO_2	28 mmHg
pO_2	126 mmHg
HCO_3^-	18 mmol/L (18 mEq/L)

Case 6. A previously well patient was brought to the emergency room in a moribund state. A chest X-ray showed pulmonary edema. The laboratory results were as follows *(6)*:

pH	7.02
pCO_2	60 mmHg
pO_2	40 mmHg
HCO_3^-	15 mmol/L (15 mEq/L)
Lactate	9.0 mmol/L (81.1 mg/dL)

Case 7. A 60-year-old man, on chronic diuretics for congestive heart failure, has a 5-days history of severe vomiting when he is admitted to the ED. His lab results are as follows:

pH	7.58
pCO_2	35 mmHg
pO_2	95 mmHg
Na	133 mmol/L (133 mEq/L)
K	4.5 mmol/L (4.5 mEq/L)
Cl	82 mmol/L (82 mEq/L)
HCO_3^-	38 mmol/L (38 mEq/L).
Anion gap	13 mmol/L (13 mEq/L)

Assessment of Case 1. The patient is in acute ventilatory failure and has acute respiratory acidosis. Bicarbonate has yet to increase and compensate to any significant degree, so she may have some degree of metabolic acidosis. The patient's oxygenation status is acceptable while breathing 50% oxygen; however, she would likely be hypoxic on room air.

Assessment of Case 2. On day 10, the patient developed metabolic alkalosis from excessive administration of bicarbonate to correct the metabolic acidosis, as indicated by the elevated lactate. Although the pCO_2 of 46 mmHg might indicate respiratory compensation, the pCO_2 probably reflects inadequate ventilation due to pulmonary edema. The pO_2 of 40 mmHg on 40% oxygen indicates severe hypoxemia. This patient may have a triple disorder of respiratory acidosis, metabolic acidosis, and metabolic alkalosis.

Revision of the central shunt dramatically improved tissue oxygenation, as determined by the rising pO_2 and decreasing lactate. As the pH normalized, pCO_2 decreased slightly to 44 mmHg, and HCO_3^- decreased to 30 mmol/L (30 mEq/L). The pCO_2 and HCO_3^- at day 15 are more consistent with a compensated respiratory acidosis. These results suggest the remnants of the metabolic alkalosis along with some degree of respiratory compensation.

Assessment of Case 3. This is likely a mixed acid−base disorder. The asthma along with several days of hyperventilation suggests that this boy has chronic respiratory alkalosis. He is severely hypoxemic and his electrolyte results show an elevated anion gap. The combination of hypoxemia with an elevated anion gap (from excess lactate) and suggests he also has a metabolic acidosis, which could explain why his pH is lower than might be expected for a compensated respiratory alkalosis. He will continue to require supplemental oxygen.

Assessment of Case 4. During the operation, the patient was in metabolic acidosis. In an alert patient, respiratory compensation would have lowered the pCO_2 more than seen here, so he appears to have some degree of respiratory acidosis due to inadequate ventilation. After the operation, while breathing room air, the patient is showing a nearly compensated metabolic acidosis, and his oxygenation status is acceptable.

Assessment of Case 5. The patient is in chronic respiratory alkalosis (chronic hyperventilation). Note that renal loss of bicarbonate has fully compensated and brought the pH to normal, as can happen in chronic respiratory alkalosis.

Assessment of Case 6. This patient has acute ventilatory failure (respiratory acidosis) from the pulmonary edema combined with metabolic acidosis from moderate to severe hypoxemia. The poor oxygenation in the lungs likely explains the decreased pO_2 of 40 mmHg and the congestive heart failure could cause poor perfusion, both of which could cause the elevated lactate. The very low pH reflects the combined respiratory and metabolic acidoses.

Assessment of Case 7. This case illustrates the importance of the clinical history in accurate interpretation of the case. This appears to be a mixed acid-base disorder with several causes. The chronic use of diuretics could cause a compensated metabolic alkalosis. From that, we would expect an elevated bicarbonate with pCO_2 elevated to normalize pH. However, the recent vomiting would also cause a metabolic alkalosis, which would elevate the pH and bicarbonate even further. This patient has a metabolic alkalosis from both vomiting and chronic diuretic use along with a respiratory alkalosis from chronic hyperventilation.

References

1. Seifter, J. L.; Chang, H.-Y. Disorders of Acid-Base Balance: A New Perspective. *Kidney Dis.* **2017,** *2* (4), 170−186.
2. Grogono, A. W. *Acid-Base Tutorial: Metabolic Acidosis and Alkalosis.* http:/www.acid-base. com. (accessed December 2008).
3. Brandis, K. *Acid-Base Physiology: Metabolic Acidosis - Causes.* From http://www. anaesthesiaMCQ.com/AcidBaseBook. (accessed January 2009).
4. Robergs, R. A.; Ghiasvand, f; Parker, D. Biochemistry of Exercise-Induced Metabolic Acidosis. *Am. J. Physiol. Regul. Integr. Comp. Physiol.* **2004,** *287,* R502−R516.
5. Kraut, J. A.; Nagami, G. T. The Serum Anion Gap in the Evaluation of Acid-Base Disorders: What Are Its Limitations and Can Its Effectiveness Be Improved? *Clin. J. Am. Soc. Nephrol.* **2013,** *8,* 2018−2024.
6. Feldman, M.; Soni, N.; Dickson, B. Influence of Hypoalbuminemia or Hyperalbuminemia on the Serum Anion Gap. *J. Lab. Clin. Med.* **2005,** *146,* 317−320.
7. Brandis, K. *Acid Base Physiology: 3.2 the Anion Gap.* www.anaesthesiamcq.com/ AcidBaseBook/ab3_2.php.
8. Hamm, L. L.; Nakhoul, N.; Hering-Smith, K. S. Acid-Base Homeostasis. *Clin. J. Am. Soc. Nephrol.* **2015,** *10,* 2232−2242.
9. Emmett, M.; Palmer, B. F. *Simple and Mixed Acid-Base Disorders,* 2020. www.uptodate. com. contents/simple-and-mixed-acid-base-disorders.
10. Burger, M.; Schaller, D. J. *Metabolic Acidosis;* NCBI Bookshelf, November 2020. https:// www.ncbi.nlm.nih.gov/books/NBK482146/.
11. Brandis, K. *Acid-Base Physiology: Metabolic Acidosis - Correction.* From http://www. anaesthesiaMCQ.com/AcidBaseBook/ab5_6.php. (accessed December 2020).
12. Brinkman, J. E.; Sharma, S. *Physiology, Metabolic Alkalosis − Stat Pearls,* 2020. www.ncbi. nlm.nih.gov/books/NBK482291/.
13. Story, D. A.; Morimatsu, H.; Bellomo, R. Strong Ions, Weak Acids and Base Excess: A Simplified Fencl-Stewart Approach to Clinical Acid-Base Disorders. *Br. J. Anaesth.* **2004,** *92,* 54−60.
14. Brandis, K. *Acid-Base Physiology: Metabolic Alkalosis - Causes.* From http://www. anaesthesiaMCQ.com/AcidBaseBook. (accessed November 2020).
15. Brandis, K. *Metabolic Alkalosis − Compensation. From Acid-Base Physiology.* http://www. anaesthesiamcq.com/AcidBaseBook/ab7_5.php. (accessed December 2020).
16. Brandis, K. *Metabolic Alkalosis − Correction. From Acid-Base Physiology: Metabolic Alkalosis.* From http://www.anaesthesiamcq.com/AcidBaseBook/ab7_6.php. (accessed December 2020).
17. Adrogue, H. J.; Madias, N. E. Management of Life-Threatening Acid-Base Disorders. *N. Engl. J. Med.* **2005,** *338,* 107−111.
18. Brandis, K. *Acid-Base Physiology: Respiratory Acidosis - Causes.* From http://www. anaesthesiamcq.com/AcidBaseBook/ab4_2.php. (accessed December 2020).
19. Patel, S.; Sharma, S. *Respiratory Acidosis;* NCBI Bookshelf, June 2020. www.ncbi.nlm.nih. gob/books/NBK482430/.
20. Narins, R. G., Ed. *Maxwell and Kleeman's Clinical Disorders of Fluid and Electrolyte Metabolism,* 5th ed.; McGraw-Hill: New York, 1994.
21. *Respiratory Acidosis − Wikipedia.* From http://en.wikipedia.org/wiki/Respiratory_acidosis.
22. Brandis, K. *Acid-Base Physiology: Respiratory Acidosis - Correction.* From http://www. anaesthesiamcq.com/AcidBaseBook/ab4_6.php. (accessed December 2020).

23. Brandis, K. *Respiratory Alkalosis — Causes*. http://www.anaesthesiamcq.com/AcidBaseBook/ab6_2.php. (accessed November 2020).

24. Brandis, K. *Bedside Rules for Assessment of Compensation*. http://www.anaesthesiaMCQ.com/AcidBaseBook/ab9_3.php. (accessed December 2020).

25. Martin, L. *Diagnosing Mixed Acid-Base Disorders*. From http://www.lakesidepress.com/pulmonary/ABG/MixedAB.htm. (accessed December 2020).

26. Walmsley, R. N.; White, G. H. Mixed Acid-Base Disorders. *Clin. Chem.* **1985**, *31*, 321–325.

27. Brandis, K. *Acid-Base Physiology: The Delta Ratio*. From http://www.anaesthesiamcq.com/AcidBaseBook/ab3_3.php. (accessed December 2020).

28. Delta Gap and Delta Ratio. Deranged Physiol. https://derangedphysiology.com/main/cicm-primary-exam/required-reading/acid-base-physiology/Chapter 705/delta-gap-and-delta-ratio. (accessed December 2020).

29. Emmett, M.; Palmer, B. *The Delta Anion Gap/Delta HCO$_3$ Ratio in Patients with a High Anion Gap Metabolic Acidosis*, 2016. www.uptodate.com. https://www.uptodate.com/contents/the-delta-anion-gap-delta-hco3-ratio-in-patients-with-a-high-anion-gap-metabolic-acidosis.

30. Schindler, E. I.; Brown, S. M.; Scott, M. G. Electrolytes and Blood Gases. In *Tietz Textbook of Clinical Chemistry and Molecular Diagnostics;* Rifai, N., Horvath, A. R., Wittwer, C. T., Eds., 6th ed.; Elsevier Saunders: St Louis, 2018; pp 604–625.

31. Klaestrup, E.; Trydal, T.; Pedersen, J.; Larsen, J.; Lundbye-Christensen, S.; Kristensen, S. Reference Intervals and Age and Gender Dependency for Arterial Blood Gases and Electrolytes in Adults. *Clin. Chem. Lab. Med.* **2011**, *49*, 1495–1500.

32. Walmsley, R. N.; White, G. H. *A Guide to Diagnostic Clincial Chemsitry*, 3rd ed.; Blackwell Scientific: Oxford, 1994; pp 79–127.

33. Kraut, J. A.; Kurtz, I. Mixed Acid–Base Disorders. In *Core Concepts in the Disorders of Fluid, Electrolytes and Acid–Base Balance;* Mount, D. B., Sayegh, M. H., Singh, A. J., Eds.; Springer: New York, 2013; pp 307–326.

34. Brandis, K. *Acid-Base Physiology: The Anion Gap*. From http://www.anaesthesiamcq.com/AcidBaseBook/ab3_2.php. (accessed December 2020).

35. Iberti, T. J.; Leibowitz, A. B.; Papadakos, P. J.; Fischer, E. P. Low Sensitivity of the Anion Gap as a Screen to Detect Hyperlactatemia in Critically Ill Patients. *Crit. Care Med.* **1990**, *18*, 275–277.

Chapter 3

Interpreting blood gas results on venous, capillary, and umbilical cord blood

Physiologic differences between arterial and venous blood for blood gas and acid—base measurements

More than any other analytes, pO_2, pCO_2, and pH change markedly from arterial to venous blood. While the pH apparently changes only slightly (i.e., 7.40 to 7.37 = 0.4% difference), the actual H ion concentration has a much greater relative change, going from 3.98×10^{-8} to 4.27×10^{-8}, a 7% increase. Arterial blood is nearly always preferred over venous blood for blood gas analysis of pH, pCO_2, and pO_2 for these reasons:

- Arterial pO_2 indicates the ability of the lungs to oxygenate, or equilibrate, the blood with alveolar air.
- Arterial blood provides an index of the oxygen and nutrients that will be provided to the tissues and cells.
- The composition of arterial blood is consistent throughout the circulatory system.

Because typical venous blood collected from an arm vein represents oxygen metabolism only in the arm, venous blood representing oxygen and carbon dioxide metabolism of a larger section of the body requires the collection of either "central" venous or "mixed" venous blood. Central venous blood is collected from a central venous catheter inserted either peripherally in the arm or directly into the subclavian or jugular vein and terminating either in the superior vena cava or the right atrium. A central venous line also allows an infusion of fluids, medications, or parenteral nutrition and monitoring of central venous pressure. Mixed venous blood is drawn from a pulmonary artery catheter that is inserted through one of the main central veins (subclavian, internal jugular, femoral) to ultimately reach the pulmonary artery. A pulmonary artery catheter samples the mixture of venous blood returning from the entire venous circulation: head and arms (via superior vena cava), the gut and lower extremities (via the inferior vena cava), and the coronary veins (via the

Blood Gases and Critical Care Testing. https://doi.org/10.1016/B978-0-323-89971-0.00010-0
53

coronary sinus) *(1)*. A pulmonary artery catheter also allows for direct measurement of pulmonary artery pressure and pulmonary capillary wedge pressure, and it can be used to estimate cardiac output. Both mixed venous and arterial blood are needed when determining parameters such as P50 and oxygen consumption (VO_2).

Interpretation of venous blood gas values

Arterial blood collected anaerobically is the standard sample for blood gas measurements to interpret acid—base and oxygenation status. Because arterial punctures are invasive, painful, and require special skills, venous blood gases are often used for the acid—base interpretation of the pH and pCO_2 results. As described elsewhere, venous blood may be collected as peripheral blood from a standard venipuncture, or as central or mixed venous blood from an existing central venous catheter or pulmonary artery catheter.

Blood gas values on arterial blood have been validated by far more research, and clinicians are more familiar with the reference intervals on arterial blood *(2)*. However, there is increased interest in utilizing venous blood gas results, especially for acid—base interpretation. The interest in venous blood gas testing comes both from the difficulties of obtaining arterial blood mentioned earlier and the availability of reliable noninvasive pulse oximetry for assessing arterial oxygenation. For several reasons, pO_2 results on venous blood are both inaccurate and have no clinical value because the oxygen has already been extracted by the tissues from the arterial blood *(3)*. A general disadvantage of venous blood is that its composition differs slightly depending on the source.

Venous blood by peripheral venipuncture is the least invasive and represents metabolism and gas exchange in the arm. Central venous blood is a mixture of venous blood from the upper half of the body *(2)*. Blood collected from a pulmonary artery catheter is the most invasive and is a mixture of venous blood returning from the entire circulatory system. Blood gas results on venous blood are best interpreted by using the venous blood gas results to estimate the arterial values, then using these estimated arterial values for clinical decisions *(3)*. For this purpose the arterial-venous differences are key. Table 3.1 lists estimated differences between arterial and venous blood:

To summarize, here are some points about the clinical utility of venous blood gas results compared to arterial results *(2)*:

- Central venous pH, pCO_2, and HCO_3 are highly correlated to similar parameters in arterial blood. On average, the central venous pH is ~ 0.03 units lower than arterial pH, and central venous pCO_2 is ~ 5 mmHg higher than arterial pCO_2,
- HCO_3 concentrations are nearly the same in central venous and arterial blood.

TABLE 3.1 Estimated corrections for converting venous results to arterial results.

	Convert central venous to arterial	Convert peripheral venous to arterial	References
pH	Add 0.03–0.05	Add 0.02–0.04 units	(3)
	Add 0.03		(4,5)
pCO_2 (mmHg)	Subtract 4–5	Subtract 3–8	(3)
	Subtract 4.5		(4,6,7)
HCO_3 (mmol/L)	Subtract 0.8		(5,7)
BE (mmol/L)	Subtract 0.3		(5,7)
pO_2 (mmHg)	Arterial higher by 63 ± 59 (cannot convert venous pO_2 results to arterial pO_2 results)		(6)

- Correcting central venous results as noted above will agree even better with values on arterial blood. Used judiciously, pH, pCO_2, and HCO_3 results on central venous blood are acceptable substitutes for similar results on arterial blood.
- In no case should pO_2 results on venous blood be used to evaluate a patient's oxygenation status. For many patients, noninvasive pulse oximetry is a sufficiently accurate substitute for arterial pO_2. However, in critical situations, pO_2 on arterial blood is required (2).

Capillary blood gases (neonatal)

Capillaries are the very small blood vessels (~ 8 μm diameter) that connect the smallest arteries (arterioles) to the smallest veins (venules) (8). In these capillaries, there is a vital gas exchange of oxygen, carbon dioxide, waste products, and essential nutrients with the tissue cells. With this exchange, there are gradients from arterial to venous blood: pH lower by 0.02–0.03; pCO_2 higher by 4–5 mmHg (0.6–0.7 kPa), and pO_2 lower by about 60 mmHg (8 kPa).

While arterial punctures or catheters provide the best sample, they are invasive and not always available. Capillary blood collection is a much less painful and safer mode of collection using a lancet or incision device that punctures the skin about 1 mm deep, making them much more appropriate for neonatal blood collections. Capillary blood from warmed heel sticks is easier and less invasive, but what about the reference values and clinical interpretation of capillary blood gas results?

As Higgins points out with his usual excellent commentary *(8)*, "capillary blood" obtained by skin puncture is a mixture of blood from punctured arterioles, capillaries, and venules, along with small contributions from interstitial and intracellular fluids.

How much are capillary blood gas results compromised?

As with reference intervals for arterial blood, reference intervals for neonatal blood gases also present practical challenges because they should be collected on healthy neonates with otherwise no need for blood gas results. As shown in Table 3.2, Cousineau et al. reported reference intervals on 118 healthy term babies between 36 and 60 h of age *(9,10)*.

A relatively old study was done on 158 paired arterial and capillary blood samples from 41 sick preterm babies between 3 h and 7 days old *(11)*. Although the values obtained would not be appropriate for reference intervals, the arterial-venous differences are useful. pCO_2 measurements in capillary blood were good approximations of pCO_2 in arterial blood, and 89% of pH values on warmed heel collections, differed by 0.05 units or less.

However, for pO_2, significant differences were observed in many cases, with 75% of results differing by 8 mmHg or more, and 53% of results differing by at least 15 mmHg. These discrepancies were always greater for unwarmed heel stick collections and the differences increased with higher arterial pO_2. However, as shown in Table 3.3, because capillary pO_2 was always lower than arterial pO_2, a normal capillary pO_2 value above, say 50 mmHg, confirms that the patient is not hypoxemic *(11)*.

As shown in Table 3.4, Cao et al. reported reference intervals for capillary blood in 23 healthy adults, as measured with the EPOC analyzer *(12)*.

TABLE 3.2 Reference intervals for neonatal capillary blood.

Parameter	Mean	Mean ± 2 SD interval
pH	7.40	7.31−7.47
pCO_2 (mmHg)	38.7	28.5−48.7
pO_2 (mmHg)	45.3	33−61
Lactate (mmol/L)	2.6	1.4−4.1
Hb (g/dL)	19.3	14.5−23.4
Glucose (mmol/L)	3.8	2.1−5.3
Ion Ca (mmol/L)	1.21	1.06−1.34

TABLE 3.3 Differences (art-cap) between arterial and capillary blood gas values in neonates.

Parameter	Mean art-cap difference	Range of art-cap differences
pH	0.001	−0.06 to +0.07
pCO_2 (mmHg)	0.4	−1.4 to +0.6
pO_2 (mmHg)	16.3	+0.8 to +45

TABLE 3.4 Reference intervals for capillary blood in adults.

Parameter	Mean	Mean + 2 SD interval
pH	7.41	7.35−7.45
pCO_2 (mmHg)	37.9	33−46
pO_2 (mmHg)	−	−
Lactate (mmol/L)	1.1	0.2−1.7
Hb (g/dL)	20.4	14.5−23.9
Glucose (mg/dL)	93	70−100
Ion Ca (mmol/L)	1.17	1.10−1.30

Umbilical cord blood gases

Blood circulation in the fetus and placenta

During pregnancy, the fetus is connected by the umbilical cord to the placenta, the organ that develops and implants in the mother's uterus during pregnancy. The placenta is essentially a life-support system for the fetus as it performs both "renal" functions and "pulmonary" gas exchange that eliminate waste products and supply oxygen and other essential nutrients to fetal blood.

There are three blood vessels in the umbilical cord: One large vein and two smaller arteries that coil around the umbilical vein. The umbilical arteries carry blood from the fetus that has a low oxygen and high CO_2 content and contains waste products from fetal metabolism. In the placenta, CO_2 and waste products diffuse across the placenta to the mother's blood, while oxygen and nutrients diffuse from the mother's blood to the fetal blood. The umbilical vein then carries this oxygenated blood from the placenta back to the fetus.

Fetal circulation is quite complex. Starting with the oxygen and nutrient-enriched blood in the umbilical vein, it flows to the baby's liver through a shunt called the ductus venosus. While a small amount of blood flows to the liver, most goes via the large inferior vena cava to the right atrium. Some blood from the right atrium goes to the right ventricle, then to the lungs (as in adult circulation). However, because the fetus has no lung function, most fetal blood bypasses the lungs and flows directly from the right atrium into the left atrium through the foramen ovale. From the left atrium, blood flows to the left ventricle where it is pumped into the ascending aorta. The ascending aorta immediately provides oxygenated blood via the carotid arteries to the fetal brain, which has the most important need for oxygen. Beyond the carotid arteries poorly oxygenated blood from the pulmonary artery enters the aorta through the ductus arteriosus and mixes with blood from the aorta to supply other organs. Some of this mixed blood goes into the umbilical arteries that carry blood to the placenta, picking up oxygen and nutrients, then flowing back to the fetus via the umbilical vein. Blood circulation in the fetus is shown in Fig. 3.1 *(13)*.

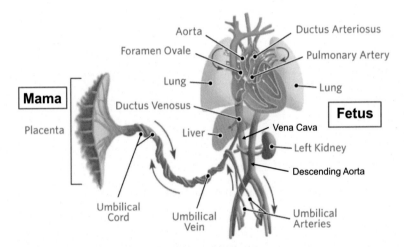

Umbilical Artery carries O_2 poor blood from fetus to placenta.
Umbilical Vein carries O_2 enriched blood from placenta to fetal circulation

https://www.chop.edu/conditions-diseases/blood-circulation-fetus-and-newborn

FIGURE 3.1 Diagram of fetal circulation. If we follow fetal circulation starting with the oxygen and nutrient-enriched blood in the umbilical vein returning from the placenta, blood flows to the fetal liver through a shunt called the ductus venosus. Most blood bypasses the liver and goes via the large inferior vena cava to the right atrium. Because the fetus has no lung function, most fetal blood bypasses the lung circulation and flows directly into the left atrium through the foramen ovale. From the left atrium, blood flows to the left ventricle where it is pumped into the ascending aorta to provide maximal oxygenated blood to the fetal brain, which has the most important need for oxygen. The other blood from the aorta mixes with the poorly oxygenated blood from the pulmonary artery entering the descending aorta to supply other organs. However, some blood from the descending aorta goes into the umbilical arteries that carry blood to the placenta, picking up oxygen and nutrients, then flowing back to the fetus via the umbilical vein. *Used with permission from the Fetal Heart Program at The Children's Hospital of Philadelphia. https://www.chop.edu/conditions-diseases/blood-circulation-fetus-and-newborn.*

Acid—base physiology in the fetus

Normal fetal metabolism produces acids that are buffered by HCO_3^- and hemoglobin to maintain blood pH within a desirable range. Most acid normally comes from the oxidative metabolism of glucose and pyruvate in the mitochondria to produce CO_2, which combines with H_2O and releases a hydrogen ion. CO_2 produced with normal metabolism is handled by the healthy fetus and placenta because CO_2 readily diffuses across the placenta. Interestingly, diffusion of CO_2 into maternal blood is facilitated by a lower maternal pCO_2 due to mild hyperventilation in pregnancy (14). When the fetus cannot remove sufficient CO_2 via the placenta, the pCO_2 increases leading to a "respiratory" acidosis. If O_2 exchange is inadequate (from multiple causes), a metabolic acidosis results.

Metabolic hydrogen ions are consumed in the synthesis of ATP from ADP, and with an adequate supply of oxygen, acid—base balance is maintained. However, with a deficit of oxygen, ADP and H^+ accumulate, causing a decreased pH. Lactate (not lactic acid) is also independently produced from a buildup of NADH that favors the conversion of pyruvate to lactate. Thus, lactate is a marker for insufficient oxygen availability to, or utilization by, the tissues, cells, and mitochondria.

Clinical uses of umbilical cord blood gas results

Hypoxia to the fetus can lead to asphyxia and acidosis during fetal development and at birth that can cause hypoxemic—ischemic encephalopathy (HIE), cerebral palsy, and harm to other organs in the fetus or newborn. Hypoxemia to the fetus may be caused by a variety of maternal, placental, and/or fetal factors (15). While transient asphyxia can occur with no pathologic consequences, more severe HIE can lead to serious neurologic deficits and death, so the challenge is to determine the severity of the oxygen debt. Umbilical cord blood gas, acid—base, and lactate measurements at birth provide more objective evidence of fetal asphyxia and distress than Apgar scores (16).

Before birth, blood may be collected by either percutaneous sampling from the umbilical cord or by fetal scalp sampling. Most commonly, samples for blood gas testing are collected from the umbilical cord soon after birth. These blood gas values provide valuable data for assessing the respiratory and metabolic status of the newborn to identify asphyxia, respiratory distress, or other conditions of fetal distress. While blood from the umbilical artery contains blood flowing from the fetus to the placenta and more accurately represents the condition of the neonate, more recent studies conclude that collection of both umbilical arterial and venous blood provides clinically useful information on the pathogenesis of acidosis (17). Indeed, some countries now require umbilical cord blood gas testing following all deliveries.

Differences in pH and pCO_2 results between umbilical arterial and venous cord blood (A—V differences) are useful for identifying fetal distress. Interpretative guidelines for A—V differences follow:

- A wide difference between umbilical arterial and venous blood results may indicate an obstructed cord.
- A small difference may be due to impaired maternal perfusion to the placenta or disruption of the maternal:placental connection that causes poor or no gas exchange at the placenta.
- A small difference may be from collecting the "arterial and venous" samples from the same umbilical vessel, most likely the umbilical vein, which has walls that are more easily punctured than the umbilical arteries.
- A pCO_2 difference between arterial and venous cord blood >25 mmHg identifies conditions such as HIE, seizures, cardiac and renal dysfunction in the neonatal period *(18)*.
- Fetal asphyxia typically results in a mixed respiratory and metabolic acidosis. When the fetus cannot remove sufficient CO_2 via the placenta, the pCO_2 increases leading to a "respiratory" acidosis. If O_2 exchange and/or delivery to fetal tissues is inadequate (from multiple causes), a metabolic acidosis develops with a further decrease in pH. Some critical results for cord blood are follows:
  1. pH less than 7.0 in arterial cord blood.
  2. When umbilical arterial pH is <7.0, the magnitude of the pCO_2 difference correlates with the risk of developing hypoxemia.
  3. A pCO_2 difference between arterial and venous cord blood >25 mmHg.
  4. A Base Deficit of ≥ 12 mmol/L *(19)*.
  5. An arterial lactate >6.7 mmol/L *(20)*.
  6. The relevance of pO_2 results on cord blood is not clearly established. However, as an indicator of an interference, one report states that an arterial umbilical cord pO_2 >38 mmHg suggests air bubble contamination in the collected blood *(21)*.

Clinical uses of umbilical cord blood lactate results

Although lactate and base deficit (BD) are related, lactate is directly measured while BD is calculated from multiple proposed equations using pH and pCO_2. Thus, lactate appears to be a more reliable indicator of outcome than pH and BD calculations *(22)*. No consensus exists yet for a critical level for arterial cord blood lactate in predicting acidosis or increased risk for a poor neonatal outcome. One report suggested that a cutoff of 8 mmol/L for arterial cord lactate indicated a higher risk for intrapartum asphyxia *(16)*. A meta-analysis

TABLE 3.5 Reference intervals for arterial and venous cord blood.

Analyte	Arterial cord blood reference range	Venous cord blood reference range	References
pH	7.14–7.42	7.22–7.44	(24)
	7.12–7.35	7.23–7.45	(20)
	7.08–7.40	7.19–7.47	(25)
pCO_2 (mmHg)	34–78	30–63	(24)
	42–74	29–53	(20)
	30–75	22–53	(25)
pO_2 (mmHg)	3–40	12–43	(24)
	6.2–28	16–40	(20)
	5–31	10–39	(25)
O_2 saturation (%)	2.7–69.0	16.4–83.3	(25)
Base deficit (mmol/L)	−1.5 to +9.3	−2.6 to +8.3	(20)
	+1.0 to +9.6	+1.0 to +7.5	(23)
Lactate (mmol/L)	2.0–6.7	−	(20)

of the literature on the use of umbilical cord lactate found that arterial umbilical cord lactate above 6.0 mmol/L had a sensitivity of 70% and a specificity of 93% for predicting poor neonatal outcomes, including HIE (23).

Reference intervals for umbilical arterial and venous cord blood

Reference intervals for arterial and venous cord blood are difficult to obtain and often vary between studies, so those from several sources are presented (20,24,25) in Table 3.5. Recall that the umbilical artery is taking less oxygenated and less nutrient-rich blood to the placenta and the umbilical vein is returning blood with a higher oxygen content and nutrient content from the placenta to the fetus.

Collecting umbilical cord blood specimens

See this in Chapter 10, *Sample collection and handling*.

References

1. Ziccardi, M. R.; Khalid, N. *Pulmonary Artery Catheterization.* NCBI Bookshelf. www.ncbi.nlm.nih.gov/books/NBK482170.

2. Higgins, C. *Central Venous Blood Gas Analysis,* July 2011. www.acutecaretesting.org.

3. Theodore, A. C.; Manaker, S.; Finlay G. *Venous Blood Gases and Other Alternatives to Arterial Blood Gases.* From www.uptodate.com. Updated 9 July 2020.

4. Malinoski, D.; Todd, S.; Slone, S.; et al. Correlation of Central Venous and Arterial Blood Gas Measurements in Mechanically Ventilated Trauma Patients. *Arch. Surg.* **2005,** *140,* 1122−1125.

5. Middleton, P.; Kelly, A.-M.; Brown, J.; et al. Agreement between Arterial and Central Venous Values for Bicarbonate Base Excess and Lactate. *Emerg. Med. J.* **2006,** *23,* 622−624.

6. Toftegaard, M.; Rees, S.; Andreassen, S. Correlation Between Acid-Base Parameters Measured in Arterial Blood and Venous Blood Sampled Peripherally form Vena Cavae Superior and from the Pulmonary Artery. *Eur. J. Emerg. Med.* **2008,** *15,* 86−91.

7. Tregor, R.; Pirouz, S.; Kamangar, N.; Corry, C. Agreement between Central Venous and Arterial Blood Gas Measurements in the Intensive Care Unit. *Clin. J. Am. Soc. Nephrol.* **2010,** *5,* 390−394.

8. Higgins, C. *Capillary-Blood Gases: To Arterialize or Not,* November 2008; pp 42−47. www.mlo-online.com.

9. Cousineau, J.; Anctil, S.; Carceller, A.; Gonthier, M.; Delvin, E. E. Neonatal Capillary Blood Gas Reference Values. *Clin. Biochem.* **2005,** *38,* 905−907.

10. Cousineau, J. *Neonate Capillary Blood Gas Reference Values,* January 2006. www.acutecaretesting.org.

11. McLain, B. I.; Evans, J.; Dear, P. R. F. Comparison of Capillary and Arterial Blood Gas Measurement in Neonates. *Arch. Dis. Child.* **1988,** *63,* 743−747.

12. Cao, J.; Edwards, R.; Chairez, J.; Devaraj, S. Validation of Capillary Blood Analysis and Capillary Testing Mode on the Epoc Point of Care system. *Pract. Lab. Med.* **2017,** *9,* 24−27.

13. https://www.chop.edu/conditions-diseases/blood-circulation-fetus-and-newborn.

14. www.uptodate.com/contents/umbilical-cord-blood-acid-base-analysis-at-delivery.

15. Higgins, C. *Umbilical-Cord Blood Gas Analysis;* Acutecaretesting.org, October 2014.

16. Gjerris, A. C.; Staer-Jensen, J.; Jorgensen, J. S.; Bergholt, T.; Nickelsen, C. Umbilical Cord Blood Lactate: A Valuable Tool in the Assessment of Fetal Metabolic Acidosis. *Eur. J. Obstet. Gynecol. Reprod. Biol.* **2008,** *139,* 16−20.

17. Jorgenson, J. S. *Umbilical-Cord Blood Gas Analysis in Obstetric Practice;* Webinar, 2015.

18. Belai, Y.; Goodwin, T. M.; Durand, M.; Greenspoon, J. S.; Paur, R. H.; Walther, F. J. Umbilical Arteriovenous pO_2 and pCO_2 Differences and Neonatal Morbidity in Term Infants with Severe Acidosis. *Am. J. Obstet. Gynecol.* **1998,** *178,* 13−19.

19. ACOG Committee Opinion # 348. Umbilical Cord Blood Gas and Acid-Base Balance. *Obstet. Anesth. Digest* **March 2007,** *27* (1), 6.

20. White, C. R. H.; Doherty, D. A.; Henderson, J. J.; Kohan, R.; Newnham, J. P.; Pennell, C. E. Benefits of Introducing Universal Cord Blood Gas and Lactate Analysis into an Obstetric Unit. *Aust. N. Z. J. Obstet. Gynaecol.* **2010,** *50,* 318−328.

21. Saneh, H.; Mendez, M. D.; Srinivasan, V. N. *Cord Blood Gas;* NCBI Bookshelf, April 20, 2020. www.ncbi.nlm.nih.gov/books/NBK545290.

22. Wiberg, N.; Kallen, K.; Herbst, A.; Olofsson, P. Relation between Umbilical Cord Blood pH, Base Deficit, Lactate, 5-minute Apgar Score and Development of Hypoxic Ischemic Encephalopathy. *Acta Obstet Gynecol. Scand.* **2010,** *89,* 1263−1269.

23. Allanson, E. R.; Waqar, T.; White, C. R. H.; Tuncalp, O.; Dickinson, J. E. Umbilical Lactate as a Measure of Acidosis and Predictor of Neonatal Risk: A Systematic Review. *Br. J. Obstet. Gynaecol.* **2016;** https://doi.org/10.1111/1471-0528.14306. www.bjog.org.

24. Fouse, B. *Reference Range Evaluation for Cord Blood Gas Parameters,* June 2002. www.bloodgas.org.

25. Arikan, G. M.; Scholz, H. S.; Petru, E.; Haeusler, M. C. H.; Haas, J.; Weiss, P. A. M. Cord Blood Oxygen Saturation in Vigorous Infants at Birth; What is Normal? *Br. J. Obstet. Gynaecol.* **2000,** *107,* 987−994.

Chapter 4

Disorders of oxygenation: hypoxemia and tissue hypoxia

Introduction

Hypoxemia, a decreased partial pressure of oxygen in the blood, is a commonly encountered clinical problem among patients with acute or chronic cardiopulmonary disorders. Accurately characterizing the cause and severity of hypoxemia in a timely manner can have significant implications for clinical management and patient prognosis. The blood gas measurements play an integral part in the evaluation of the hypoxemic patient, as does the understanding of oxygen uptake, delivery, and consumption.

A small amount of oxygen is dissolved in the blood and the rest is bound to hemoglobin (Hb). Effective utilization of oxygen requires binding by Hb in the lungs, transport in the blood vessels, and release to tissues, where cellular respiration occurs. Hydrogen ion (pH), CO_2, temperature, and 2,3-DPG all play important roles in these processes.

A panel of blood gas measurements includes partial pressures of oxygen (pO_2) and carbon dioxide (pCO_2), percent oxyhemoglobin ($\%O_2Hb$), and related parameters, which are used to detect and monitor oxygen deficits from a variety of causes. Measurement of blood gases and cooximetry may be done by laboratory analyzers, point-of-care testing, and transcutaneous blood gas devices.

Identifying the cause of an oxygen deficit requires an understanding of the measurements and calculations important in oxygen uptake and delivery, which include the Hb concentration, alveolar-arterial pO_2 (A-a) gradient, pO_2:FiO_2 ratio, oxygenation index (OI), O_2 content, O_2 delivery, pulmonary dead space, and intrapulmonary shunt.

Oxygen measurements are commonly done on arterial blood but are also done on capillary, venous, and mixed venous blood. It is important for both clinicians and laboratorians to understand the benefits and limitations that are inherent to the different measuring techniques and sample sources. Whether monitoring oxygenation is more beneficial when done in a clinical laboratory or at the point of care depends on a combination of factors, including clinical need, therapeutic urgency, proximity of the patient care areas to the clinical laboratory, and qualifications/cooperation of the nonlaboratorian testing personnel.

Blood Gases and Critical Care Testing. https://doi.org/10.1016/B978-0-323-89971-0.00012-4

65

Parameters in oxygen monitoring

The oxygen-related parameters measured by a blood gas/cooximetry analyzer are pO_2, $\%O_2Hb$, O_2 saturation (sO_2 or "O_2 sat"), and total Hb (and hematocrit). There are subtle but important differences between these individual measurements, which are discussed later in this chapter. Cooximeters measure several common forms of Hb in blood: oxyhemoglobin (O_2Hb), deoxyhemoglobin (HHb), carboxyhemoglobin (COHb), and methemoglobin (metHb).

The pO_2 represents the tension of oxygen in the aqueous or plasma phase of blood. pO_2 is reported in either mmHg in the United States or kPa in most other countries. The typical atmospheric pressure at sea level is 760 mmHg (101.3 kPa) and the atmosphere contains 21% oxygen. Thus, the pO_2 of inspired air at sea level is 160 mmHg (21.3 kPa). If separate open containers of blood and water are exposed to the atmosphere, the pO_2 should be the same for each, approximately 160 mmHg. However, the blood is not directly exposed to ambient atmospheric air. As the air is inspired through the nasopharynx and travels down the trachea and bronchi, it is humidified and the partial pressure of water increases from approximately 3 mmHg in ambient air to 47 mmHg at the level of the alveoli. Because of CO_2 released from the blood, the partial pressure of CO_2 in the alveolar space increases from 0.3 mmHg to approximately 40 mmHg in healthy individuals at rest. The presence of increased H_2O and CO_2 in the alveolar space displaces pO_2 and leads to an alveolar pO_2 of approximately 104 mmHg. Thus, arterial pO_2 should be close to equilibrating with alveolar pO_2 in normal adults with only a small decrease accounted for by the transport across the alveolar-capillary membrane.

The amount of dissolved oxygen in the blood, as measured by the pO_2 of the sample, contributes only a small amount to the overall oxygen content of the blood. The majority of the oxygen carried in the blood is bound to Hb. The amount of oxygen bound to Hb may be represented by either the percent oxygen saturation (sO_2) or the percent oxyhemoglobin ($\%O_2Hb$). The sO_2 is a measurement of the percentage of the O_2Hb relative to the total *functional* Hb ($O_2Hb + HHb$). It is calculated as follows:

$$sO_2 = O_2Hb/(O_2Hb + HHb)$$

If all Hb binding sites contain oxygen, then the sO_2 would be 100%, although the normal range for sO_2 is typically 94%−98%.

The $\%O_2Hb$ is the percent of the *total* Hb that contains oxygen. It is calculated as follows:

$$\%O_2Hb = O_2Hb/(O_2Hb + HHb + COHb + metHb)$$

Because blood usually contains some COHb and metHb, the $\%O_2Hb$ never reaches 100% and typically ranges from about 92% to 96% in healthy individuals, slightly lower than the sO_2. In patients with elevated COHb or metHb, the $\%O_2Hb$ will be decreased, but the sO_2 is unaffected.

The pO_2 and the sO_2 are related as illustrated in the oxyhemoglobin dissociation curve (Fig. 4.1). As the pO_2 increases up to about 100 mmHg, the sO_2 also increases to about 99%−100% (i.e., all the hemoglobin oxygen binding sites contain oxygen). However, if a patient breathes supplemental oxygen (such as an FiO_2 of 0.40), the arterial pO_2 may increase well above 100 mmHg, while the sO_2 remains at or near 100%.

The hematocrit is the percentage of the blood volume that is occupied by RBCs. The hematocrit is directly proportional to the hemoglobin concentration. The numerical hematocrit percent is generally about three times the hemoglobin concentration (in g/dL). The oxygen content is directly proportional to the sO_2 percentage and the hemoglobin concentration (see *"Measured and Calculated Parameters Used in Evaluating Arterial Oxygenation"*). The oxygen content and the cardiac output are used to calculate the oxygen delivery.

Measurement of oxygen levels (i.e., pO_2 and $\%O_2Hb$) in peripheral venous blood is of limited utility in assessing a patient's oxygenation status; however, pH and pCO_2 measurements can still be quite useful in estimating a patient's acid−base status and level of ventilation. Mixed venous blood samples obtained from the pulmonary artery after all the blood returned from the body has been mixed and pumped into the pulmonary circulation from the right ventricle can give an estimate of oxygen consumption. For example, in the

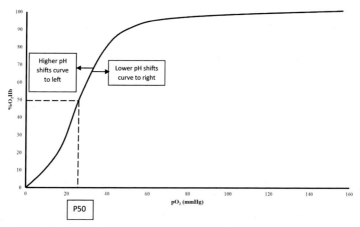

FIGURE 4.1 Oxyhemoglobin dissociation curve for whole blood. The line represents the percent of Hb that becomes saturated with oxygen as the pO_2 increases. As the pO_2 increases up to about 100 mmHg, the sO_2 also increases to about 99%−100%. If a patient breathes supplemental oxygen, the arterial pO_2 may increase much higher (i.e., 200−400 mmHg), but the sO_2 cannot increase beyond 100%. The P50 is the pO_2 at which the $\%O_2Hb$ is 50%. An increase in pH shifts the curve to the left while a decrease in pH shifts the curve to the right.

setting of sepsis, patients may have a high cardiac output, mitochondrial dysfunction and poor oxygen extraction. Therefore, they will have high mixed venous %O_2Hb, but this does not necessarily indicate that their tissue oxygen uptake is adequate. In patients with heart failure and low cardiac output, they will often have insufficient oxygen delivery but normal capacity for oxygen extraction and subsequently a low mixed venous %O_2Hb. These two conditions are frequently present together and can complicate the interpretation of mixed venous %O_2Hb and therefore limit its clinical utility.

Pulse oximetry has become universally utilized for continuously and noninvasively monitoring a patient's arterial hemoglobin saturation without needing to invasively measure sO_2 with a blood gas analyzer. In the absence of more severe hypoxemia (sO_2 <90%) and states of poor perfusion, the measurements obtained with a pulse oximeter generally approximate blood sO_2 with good accuracy (1). Because pulse oximetry will not measure the oxygen saturation correctly for other hemoglobins such as metHb or COHb, pulse oximetry will not detect CO poisoning. For example, if a patient has an elevated COHb, the %O_2Hb will be decreased, but the pulse oximeter may read a normal sO_2. A cooximeter, which requires a blood sample, will determine the %O_2Hb (and sO_2 if desired), and the percentages of metHb and COHb.

Point-of-care testing for blood gases. Point-of-care testing (POCT) has the potential to deliver rapid blood gas and other critical care test results near the patient. When properly incorporated into the patient care process, POCT can translate to faster therapeutic intervention, reduced preanalytical errors, and improved patient care. However, POCT also requires a higher level of supervision and quality management to avoid the pitfalls of improper sample handling, test inaccuracy, training and continued competency assessment of nonlaboratorians, and justification of the additional costs of analyzers and test units or cartridges. The benefits of POCT are also dependent on the test volume and proximity of the care area to a clinical laboratory. For example, POCT becomes very labor intensive for high volume testing and may have limited benefit if a clinical laboratory is located near the patient-care area.

Transcutaneous monitoring. Transcutaneous monitors for pO_2 (and pCO_2) measure the pO_2 at the skin surface ($p_{tc}O_2$), which gives an estimate of the arterial pO_2. Monitors typically warm the skin at the sensor site to increase blood flow. Since pulse oximetry has become a standard of care for noninvasive monitoring of oxygen levels, the $p_{tc}O_2$ is used only infrequently and predominantly in neonates. When initiating transcutaneous monitoring, it is recommended that an arterial blood gas be drawn and measured for pO_2 to ensure accurate calibration. Therefore, the clinical utility of transcutaneous monitoring is limited to a small subset of patients who either lack an arterial access site or require continuous monitoring of oxygen and carbon dioxide with minimal blood draws.

Structure and function of hemoglobin

The four most prevalent forms of normal hemoglobin in blood are oxyhemoglobin (O_2Hb), deoxyhemoglobin (HHb), carboxyhemoglobin (COHb), and methemoglobin (metHb). O_2Hb and HHb are both functional hemoglobin and represent the same molecules as they continually bind and release O_2. Both COHb and metHb are nonfunctional in O_2 binding and release.

Four subunits join together to form the hemoglobin molecule (Fig. 4.2). Each subunit contains a globin protein + heme + Fe^{++} that binds O_2. Both O_2 and CO_2 (carbaminohemoglobin) bind *reversibly* to hemoglobin. As one O_2 binds to a subunit, the conformation of the remaining subunits changes slightly to favor more O_2 binding. This is positive cooperativity.

There are several molecules, ions, and conditions that promote the timely binding of oxygen to hemoglobin in the lungs and release in the tissues and cells (Table 4.1). These are hydrogen ions (H^+), measured as its negative log pH, pCO_2, temperature, and 2,3-DPG. Conditions of higher metabolic activity or oxygen consumption, as in cells, tend to produce higher $[H^+]$ (acidity), higher pCO_2, and warmer temperatures that favor the release of O_2 from Hb, while less $[H^+]$ (alkalinity), lower pCO_2, and cooler temperature, as in the lungs, tend to favor binding of O_2 to Hb.

The role of 2,3-DPG on Hb binding and release of O_2 is more complex. 2,3-DPG is an ion largely contained in RBCs that "cooperates" with O_2 and other molecules to control when Hb binds or releases O_2. Without any 2,3-DPG, Hb would stay in a conformation (R form) with a higher affinity for O_2 that favors the binding of O_2. As blood enters the tissues, the increased

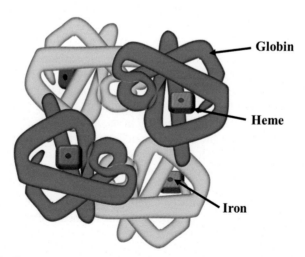

FIGURE 4.2 **Structure of the hemoglobin molecule.** Four subunits join together to form the hemoglobin molecule. Each subunit contains a globin protein + heme + Fe^{++} that binds O_2.

TABLE 4.1 Factors that favor oxygen binding to or release from hemoglobin (Hb).

Favors O_2 binding to Hb	Favors O_2 release from Hb
↑ pH	↓ pH
↓ pCO_2	↑ pCO_2
↓ temperature	↑ temperature
↓ 2,3-DPG bound to Hb	↑ 2,3-DPG bound to Hb

H^+ ions and pCO_2 promote the release of O_2, which then favors the binding of 2,3-DPG to Hb. This causes a conformational change to the T form of Hb that has a lower affinity for O_2, therefore leading to further release of oxygen to the tissues.

As blood enters the lungs, which have a higher pO_2, as some O_2 binds to Hb, more 2,3-DPG is released, which then promotes more binding of O_2 to Hb *(2)*.

As a note, fetal Hb does not bind 2,3-DPG as well as adult Hb, so fetal Hb tends to stay in a conformation with a higher affinity for O_2 than does adult Hb. This is essential so that the fetal blood can more effectively extract O_2 from maternal blood.

Carbon monoxide is small enough to fit into the protein crevice and create very strong bonds with iron to form COHb, which is unable to release oxygen. The normal level of COHb in the blood is about 1%, but in heavy smokers COHb levels may be up to 10% of the total Hb. Because hemoglobin binding affinity for CO is 200 times greater than its affinity for oxygen, even when inspired air contains CO levels as low as 0.02%, headache and nausea occur *(3,4)*. If the CO concentration increases to 0.1%, unconsciousness will follow. While COHb is nonfunctional, if CO can be removed, the hemoglobin becomes functional again. CO releases very slowly from COHb once the source of CO (i.e., cigarette smoke) has been eliminated, but this process can be accelerated in the presence of higher pO_2 in the blood obtained by providing either supplemental oxygen or hyperbaric oxygen to the patient. In CO poisoning, the affected person dies of asphyxiation because their blood is no longer able to carry enough oxygen to supply the needs of the tissues and brain.

MetHb has its Fe^{++} ions oxidized to Fe^{+++}, which renders the hemoglobin nonfunctional in oxygen transport *(2)*. Methemoglobinemia can be congenital, due to genetic mutations in the enzyme cytochrome b5 reductase, which normally maintains the balance of metHb by reducing Fe^{+++} back to Fe^{++} forming HHb, or acquired, typically secondary to medications or exogenous chemicals. At metHb levels above 10%, symptoms due to tissue hypoxia may range from mild cyanosis, dyspnea, headache, or fatigue. As metHb levels progressively increase above 20%, hypotension, cardiac dysfunction, seizures, coma, and even death may occur at metHb levels above 50%. Treatment of severe, symptomatic methemoglobinemia involves removing any offending drugs and administering methylene blue or ascorbic acid to facilitate reduction of metHb back to HHb.

Another poisonous agent that binds to hemoglobin is the cyanide anion (CN^-). This toxic agent is inhaled as hydrogen cyanide gas. Once cyanide is taken into the bloodstream the majority (92%—99%) binds to hemoglobin in red blood cells. From there, it is taken to the cells and mitochondria where it binds to the enzyme cytochrome oxidase and inhibits mitochondria from metabolizing oxygen.

Processes in oxygen transport and delivery to tissues and mitochondria

Oxygen is utilized at the level of the mitochondria to generate ATP, which supplies molecular energy needed for many of the cellular processes vital to human life. The full benefit of oxygen is only realized when all of the steps from initial inspiration of oxygen through the nose and mouth to oxidative phosphorylation in the mitochondria are working effectively. After recognizing the patient may have an inadequate supply or utilization of oxygen, the clinician is faced with determining which process or processes are dysfunctional and what are the most appropriate therapies. We will further explore these processes (listed below) in the following sections.

- Air intake by spontaneous breathing or mechanical ventilation
- Air entry into the alveoli
- Oxygen diffusion across the alveolar-capillary membrane
- Oxygenation of blood/hemoglobin
- Oxygen transport to tissues and cells
- Oxygen release from Hb and diffusion into cells
- Oxygen utilization by the mitochondria to produce adequate ATP

Evaluation of hypoxemia

Hypoxemia is a commonly encountered clinical abnormality in both the inpatient hospital setting and the outpatient clinic. Clinicians must be able to quickly evaluate a patient to determine the etiology of hypoxemia and potentially mitigate its ill effects. We will discuss the five physiologic mechanisms through which hypoxemia occurs, and how to determine the presence of each one (Table 4.2).

TABLE 4.2 Effects of hypoxemic mechanisms on Alveolar-arterial pO_2 gradient, chest X-ray, and effectiveness of supplemental oxygen therapy.

Mechanisms of hypoxemia	A-a gradient	Chest X-ray	Resolves with supplemental O_2
Low inspired O_2 tension	Normal	Normal	Yes
Hypoventilation	Normal	Normal	Yes
V/Q mismatch	Abnormal	Either	Yes
Right-to-left shunt	Abnormal	Either	No
Diffusion impairment	Abnormal	Abnormal	Yes

Low inspired O_2 tension. Examples of conditions of low inspired O_2 include higher altitudes with less oxygen rich air, fire, where oxygen is being consumed, or mixed gases, such as improperly calibrated inhaled anesthetics used in the operating room. These problems should be readily apparent to the clinician based on the setting and rapidly improve with the administration of supplemental oxygen.

Hypoventilation. Hypoventilation occurs when a patient's alveolar ventilation is inadequate to sufficiently clear CO_2 from the lungs, which also increases blood pCO_2 (hypercarbia). This can be seen in obstructive lung diseases (i.e., chronic obstructive lung disease or asthma), use of certain medications or illicit drugs that can blunt the respiratory drive (i.e., morphine, heroin, propofol, or benzodiazepines), or obesity hypoventilation syndrome. The elevated alveolar pCO_2 displaces some of the inspired oxygen, effectively lowering the alveolar oxygen tension. Hypoxemia secondary to hypoventilation in a hypercarbic patient can be identified by calculating an alveolar-arterial oxygen gradient (A-a gradient, as described in the next section), which takes into account the elevated pCO_2 and should be near normal in cases of hypoxemia caused by pure hypoventilation.

V/Q mismatch. The term V/Q (Ventilation/Perfusion) refers to the amount of air entering the alveoli (V) relative to the capillary perfusion (Q) of those alveoli.

- A V/Q ratio of 1.0 means that the ventilation to an alveolar unit (such as 1 mL/min) is available to exchange gases with an equal amount of alveolar capillary blood (1 mL/min).
- A V/Q ratio of 2.0 indicates twice as much alveolar ventilation as alveolar capillary perfusion. A simple example of a condition leading to a V/Q mismatch of 2.0 would be a pulmonary embolism where a segment of the lung receives ventilation, but the blood flow to that ventilated area is blocked by a blood clot. Some degree of V/Q mismatching is present with

most acute and chronic pulmonary diseases. Under most circumstances, hypoxemia secondary to V/Q mismatch can be overcome by administering supplemental oxygen.

- A ratio of 0.5 indicates half as much ventilation as perfusion *(5)*, an example of which is shunting described next.

Right-to-Left shunt. A shunt indicates that blood goes from the right side of the heart to the left side of the heart without coming in contact with functioning alveoli. While a small amount of shunting is normal, this can occur pathologically due to anatomic intracardiac or vascular abnormalities, but most commonly occurs due to intrapulmonary shunting. An intrapulmonary shunt represents extreme V/Q mismatching with a V/Q ratio of 0 for a given segment of the lung. This can occur in the setting of atelectasis, pneumonia, or complete bronchial obstruction where pulmonary blood perfuses nonfunctional or nonventilated alveoli. Hypoxemia secondary to a right-to-left shunt cannot be fully overcome by administering supplemental oxygen.

Diffusion impairment. Conditions that increase the distance between alveolar gas and capillary blood lead to diffusion impairment. This can occur when the alveoli are partially filled with fluid, hyaline membranes, or other debris, which can be seen in infectious pneumonia or acute respiratory distress syndrome. Other causes of diffusion impairment are diseases that cause thickening of the interstitium, or area between alveolar epithelium and capillary endothelium, and are broadly called interstitial lung diseases. Hypoxemia caused by a condition leading to diffusion impairment should be apparent on clinical examination and/or chest radiograph, and it can be overcome with supplemental oxygen.

Measured and calculated parameters for evaluating arterial oxygenation

A number of oxygen parameters are used clinically to help understand the cause and/or severity of hypoxemic respiratory failure, therefore allowing physicians to develop a plan of care and/or define a patient's prognosis.

Alveolar-arterial pO_2 gradient. The A-a gradient is a measure of the difference of pO_2 between the alveoli (p_AO_2) and the arterial blood (p_aO_2). It is useful in determining both the etiology and the severity of hypoxemia *(4–6)*.

Alveolar pO_2 (p_AO_2) calculations require that we know the atmospheric pressure (760 mmHg at sea level), FiO_2 (0.21 at room air), and arterial pCO_2 (mmHg):

$$\text{Alveolar } pO_2(p_AO_2) = A = [p_{atm} - p_AH_2O] \times FiO_2 - p_aCO_2/0.8$$

- The partial pressure of water vapor in alveoli is 47 mmHg
- 0.8 is the respiratory quotient. It is the assumed ratio of (measured) arterial pCO_2 to alveolar pCO_2. For example, if the arterial pCO_2 is 40 mmHg, the alveolar pCO_2 would be 50 mmHg.

Thus, at sea level and room air:

$$A = [760 - 47] \times 0.21 - p_aCO_2/0.8$$

$$A = 150 - p_aCO_2/0.8$$

For a person breathing room air:

$$(A - a)\text{gradient} = 150 - p_aCO_2/0.8 - p_aO_2$$

The A-a gradient is helpful in determining whether a patient with an elevated p_aCO_2 has a secondary cause of hypoxemia, or whether it is due solely to reduced alveolar ventilation (Table 4.3). A low p_aO_2 with a normal A-a gradient implies that lung gas exchange function is good, but respiratory effort may be reduced, such as by reduced minute volume, obstruction of airflow, or disorders of lung structure and air flow (increased dead space ventilation).

A low p_aO_2 with an elevated A-a gradient indicates poor gas exchange. This may be caused by edema, inflammation, fibrosis, damage to alveolar capillaries, an imbalance in ventilation to perfusion (V/Q), or right-to-left shunting.

p_aO_2:FiO₂ ratio. The ratio of the p_aO_2 to FiO_2 is a simple, readily available clinical calculation used as one of the criteria to determine the presence and severity of acute respiratory distress syndrome (ARDS). A normal p_aO_2/FiO_2 ratio of approximately 400−500 represents a p_aO_2 of approximately 80−100 while breathing an FiO_2 of 0.21. A lower ratio indicates worsened lung function, more severe disease, and increased risk of mortality in ARDS (Table 4.4) (7,8).

TABLE 4.3 Calculated alveolar-arterial (A-a) gradient for normal oxygenation, oxygenation failure, and ventilation failure.

Parameter (mmHg)	Normal conditions	Oxygenation failure	Ventilation failure
pO_2 of inspired air	150	150	150
pO_{2art}	90	45	45
pCO_{2art}	40	40	75
Alveolar pO_2 (p_AO_2)	100	100	56
A-a difference	10	55	11

Oxygenation failure is when alveoli are not able to adequately equilibrate blood with alveolar air; *Ventilation failure* is when lungs are unable to inhale an adequate volume of air.

TABLE 4.4 Severity of acute respiratory distress syndrome (ARDS) related to p_aO_2/FiO_2 ratio and mortality.

Severity of ARDS	p_aO_2/FiO_2 ratio	Mortality (%)
Mild	201–300	27–35
Moderate	101–200	32–40
Severe	≤100	45–46

ARDS is a disorder characterized by diffuse damage to the alveolar-capillary membrane leading to noncardiogenic pulmonary edema and hypoxemia. The most common causes of ARDS include pneumonia, sepsis, aspiration, trauma, and pancreatitis, which lead to a systemic inflammatory response and injury to the alveolar-capillary membrane. The widespread increase in capillary permeability results in exudation of protein-rich fluid into the interstitium and alveoli.

ARDS is clinically defined as an acute onset of hypoxemic respiratory failure, a p_aO_2/FiO_2 ratio less than 300 (i.e., A patient with an FiO_2 of 0.50 who has a p_aO_2 that is less than 150 mmHg), and bilateral infiltrates on chest X-ray that are not due to cardiogenic pulmonary edema.

Oxygenation index (OI). The OI is another calculation used to determine the severity of hypoxemia in patients with ARDS *(9)*. In mechanically ventilated patients, the OI takes into account the level of pressure applied by the ventilator, which can affect the number of functional alveoli that are ventilated and therefore impact oxygenation. To calculate OI, you need the FiO_2 *(%)*, the arterial pO_2 (mmHg), and the mean airway pressure (MAP; cm H_2O):

$$OI = (FiO_2 \times MAP)/p_aO_2$$

A lower OI represents less severe hypoxemia, because this indicates lower levels of FiO_2 and ventilator pressure are required to achieve adequate oxygenation. Note that the MAP is not a laboratory measurement, but instead determined by the settings on the mechanical ventilator and the compliance of the lungs. Lungs that are severely injured often require a much higher MAP to maintain aeration.

O_2 content. The oxygen content of blood is the total concentration of oxygen in the blood, both that bound to Hb and that dissolved in the plasma *(4)*. It is commonly reported as mL O_2 per 100 mL of blood, and is sometimes called the *volume % O_2*. To calculate the O_2 content, we need the total Hb (g/dL), %O_2Hb, and the p_aO_2 (mmHg):

$$O_2 \text{ content} = (1.36 \text{ mL } O_2/\text{g Hb} \times [Hb] \times \%O_2Hb)$$
$$+ (p_aO_2 \times 0.003 \text{ mL } O_2/\text{mmHg/dL})$$

Some confusion may arise with the units used for O_2 content. The factor 1.36 represents the gas volume (mL) of oxygen bound by each gram of Hb if that O_2 gas is released from Hb at standard temperature and pressure. It has evolved from the original work of van Slyke, who measured the O_2 content of blood by chemically releasing the O_2 gas from Hb and measuring the gas volume *(10)*.

It is important to note that the hemoglobin concentration is the biggest factor in determining O_2 content so that patients with anemia (a reduction in the oxygen carrying capacity of the blood due to reduced hemoglobin or erythrocyte concentrations) will have a much lower O_2 content.

As one can see from the small conversion term of 0.003, the p_aO_2 contributes little (about 2%) to the O_2 content of blood in a person breathing room (atmospheric) air. For a person breathing oxygen-enriched air, the contribution becomes more significant, with the p_aO_2 contributing up to 8%–10% of the oxygen content for a person breathing air with a p_aO_2 of 400–500 mmHg.

O_2 delivery. Oxygen delivery (DO_2) is a product of the cardiac output (***CO***) and the arterial oxygen content (C_aO_2) and is a measure of the volume of oxygen delivered to the systemic circulation each minute. It is the mL/minute of O_2 pumped by the heart to the body.

$$DO_2 = CO \times C_aO_2$$

Pulmonary dead space. Pulmonary "dead space" is the gas volume in the lungs that does not undergo gas exchange. This is composed of both the anatomic dead space and alveolar dead space. Normal anatomic dead space comprises the conducting airways and represents approximately 30% of the volume of inhaled air during normal breathing at rest. Alveolar dead space represents unperfused or underperfused alveoli, as with diseases such as emphysema, COPD, and pulmonary emboli that diminish blood flow to alveoli.

The total pulmonary dead space fraction represents the fraction of the tidal volume that does not reach functioning alveoli:

Dead Space Fraction = (Tidal Volume − Alveolar ventilation)/Tidal Volume

While the anatomic dead space is important, the alveolar dead space is the most relevant to disease states. Alveolar dead space can be calculated using the Bohr Equation and can be performed at the bedside on a mechanically ventilated patient if an end-tidal CO_2 monitor and arterial blood gas are available.

$$\text{Alveolar dead space fraction} = \frac{p_aCO_2 - p_{ET}CO_2}{p_aCO_2}$$

where $p_{ET}CO_2$ is the end-tidal pCO_2 of exhaled air.

Right-to-left shunt fraction. Intrapulmonary shunting occurs both normally and pathologically. Normally, approximately 2%–5% of cardiac output perfuses the lungs to supply oxygen to the bronchial cells and pleura and returns to the left atrium without coming in contact with the alveolar capillaries. Pathologic intrapulmonary shunting occurs when blood perfuses nonfunctional or nonventilated alveoli. This may be calculated as the *shunt fraction*, defined as the ratio of shunted cardiac output to the total cardiac output.

The shunt fraction is a more difficult parameter to calculate because it requires both an arterial and a mixed venous blood sample:

$$\frac{\text{Shunted cardiac output}}{\text{Total cardiac output}} = \frac{Q_S}{Q_T} = \frac{C_{PC}O_2 - C_aO_2}{C_{PC}O_2 - C_vO_2}$$

$C_{PC}O_2$ is the oxygen content of pulmonary capillary blood, which should have the highest O_2 content in the circulation. For calculations, it is assumed this blood is 100% saturated and has a pO_2 of 101 mmHg:

$$C_{PC}O_2 = (1.36 \times Hb) + (101 \times 0.003)$$

C_vO_2 is the oxygen content of a mixed venous blood sample taken from the pulmonary artery.

Methodology of oxygen measurements

pO_2 **Electrodes.** A typical pO_2 electrode consists of a silver anode, platinum cathode, and Ag/AgCl reference electrode, all covered by a thin film of electrolyte solution and polypropylene membrane that is permeable to oxygen gas but restricts proteins and other potential contaminants (Fig. 4.3). Oxygen from the sample diffuses across the polypropylene membrane into the electrolyte solution and is reduced on the cathode (electrons are consumed) according to this equation:

$$O_2 + 4H^+ + 4e^- \rightarrow 2H_2O \, (\text{the } H^+ \text{ ions come from the electrolyte solution})$$

The flow of electrons produces a current that is proportional to the amount of O_2 present in the sample. To complete the electrical circuit, Ag is oxidized at the anode. The amperometric current measured during this process is automatically converted to a $pO2$ value by the analyzer.

This relationship is described by the equation:

$$pO_2 = (I - I_o)/S$$

I is the electrode current measured for the sample, I_o is the zero or baseline current, and S is the sensitivity of the electrode response. The values of S and I_o are calculated from calibration data for the sensor. When the electrode current I is measured for the individual blood sample being analyzed, the equation can then be solved for pO_2 *(11).*

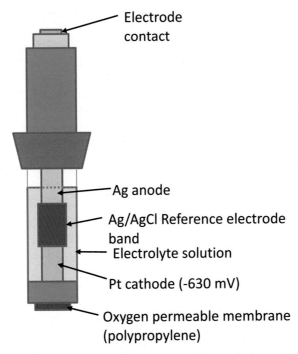

FIGURE 4.3 *p*O$_2$ **electrode.** The *p*O$_2$ electrode has a platinum cathode, a silver anode, and an Ag/AgCl reference electrode. The platinum cathode is polarized by applying a slightly negative voltage of approximately 630 mV. The electrode is capped by an O$_2$ permeable polypropylene membrane. O$_2$ migrates to the cathode and is reduced, picking up electrons. The Ag anode functions by oxidizing Ag to Ag$^+$ and producing an electron to complete the electrical circuit. Changes in the current flowing between the anode and cathode result from the amount of O$_2$ reduced in the electrolyte solution and are proportional to the partial pressure of O$_2$ in the blood sample being analyzed.

Cooximety. Modern cooximeters are often based on a spectrophotometric optical system that monitors over 100 wavelengths in the absorbance spectra of O$_2$Hb, HHb, COHb, and metHb, over a range of 478−672 nm. The blood sample may or may not be hemolyzed before being transported to the cuvette that is regulated at a temperature of 37°C. If a hemolyzer is used, the sample is ultrasonically hemolyzed at a frequency of about 30 kHz, which ruptures the red cell membranes. Light from a source (halogen lamp or LED) is transmitted through the cuvette passing through a slit to a grating where the light is separated into multiple single wavelengths that are each monitored by a photodiode array. The currents and therefore the intensity of the light signals are measured at each of the 100-plus diodes, which form the basis for the absorption spectrum for a particular sample. The spectrum is analyzed and the oximetry parameters (O$_2$Hb, COHb, metHb, HHb, and total Hb) are calculated and added together to give the total Hb concentration.

Pulse Oximetry. Pulse oximetry uses light absorption through a pulsing capillary bed usually in a toe or finger. The probe uses two LED light sources; one is red (660 nm) and the other is invisible infrared ($\sim$940 nm). Although some light is absorbed by skin and tissue, the only variable absorption is due to arterial pulsations. These absorbance differences at different wavelengths are used to calculate sO_2 for hemoglobin.

Self-assessment and mastery

Self-assessment questions

1. Which of the following represents normal alveolar gas at sea level in the normal human lung at rest?
   a. $pN_2 = 566$ mmHg, $pO_2 = 120$ mmHg, $pCO_2 = 27$ mmHg, $pH_2O = 47$ mmHg
   b. $pN_2 = 569$ mmHg, $pO_2 = 104$ mmHg, $pCO_2 = 40$ mmHg, $pH_2O = 47$ mmHg
   c. $pN_2 = 273$ mmHg, $pO_2 = 400$ mmHg, $pCO_2 = 40$ mmHg, $pH_2O = 47$ mmHg
   d. $pN_2 = 564$ mmHg, $pO_2 = 149$ mmHg, $pCO_2 = 0.3$ mmHg, $pH_2O = 47$ mmHg
2. Which of the following equations represents the percent oxygen saturation (sO_2)?
   a. $(1.36 \text{ mL } O_2/\text{g Hb} \times [\text{Hb}] \times \%O_2\text{Hb}) + (p_aO_2 \times 0.003 \text{ mL } O_2/\text{mmHg/dL})$
   b. $[p_{atm} - p_AH_2O] \times FiO_2 - p_aCO_2/0.8$
   c. $O_2\text{Hb}/(O_2\text{Hb} + \text{HHb} + \text{COHb} + \text{metHb})$
   d. $O_2\text{Hb}/(O_2\text{Hb} + \text{HHb})$
3. What is the oxygen content of blood with a $pO_2 = 80$ mmHg, $pCO_2 = 65$ mmHg, Hb = 9.5 g/dL, $\%O_2\text{Hb} = 95\%$ in a patient breathing 100% oxygen?
   a. 12.5 mL/dL
   b. 80
   c. 613 mmHg
   d. 533
4. Which of the following enhances O_2 release from hemoglobin?
   a. carbon monoxide
   b. alkalemic pH (>7.45)
   c. fever (T > 38 C)
   d. low pCO_2 (<30 mmHg)
5. Which of the following is not a physiologic cause of hypoxemia?
   a. anemia
   b. V/Q mismatch
   c. hypoventilation
   d. right-to-left shunt

6. Which p_aO_2/FiO_2 ratio in a patient with ARDS predicts the highest mortality?
   a. 50
   b. 175
   c. 200
   d. 300

Self-assessment questions answer key

1. **b.** Choice a represents the gas composition of mixed expired air, which has a slightly higher pO_2 and lower pCO_2 because it represents expired alveolar air mixed with the air occupying the conducting airways of the anatomic dead space. Choice d represents the composition of inspired air in the conducting airways.
2. **d.** Choice a is the equation for oxygen content of the blood, choice b is the equation for alveolar oxygen content (p_AO_2), and choice c is the equation for percent oxyhemoglobin (%O_2Hb).
3. **a.** Choice b is the p_aO_2/FiO_2 ratio, choice c is the p_AO_2, and choice d is the alveolar-arterial difference (A-a gradient) based on the patient's values provided.
4. **c.** Conditions of higher [H^+], higher pCO_2, higher temperature, and greater concentration of 2,3-DPG favor O_2 release from hemoglobin, whereas the opposite of these conditions, along with the presence of some toxins, such as carbon monoxide tend to favor stronger binding of O_2 to hemoglobin.
5. **a.** The five physiologic causes of hypoxemia (a low arterial pO_2), are low inspired oxygen tension, hypoventilation, V/Q mismatch, right-to-left shunt, and diffusion impairment. While anemia will decrease the oxygen content of the blood by having less hemoglobin binding to oxygen, the arterial pO_2 will not be impacted by anemia.
6. **a.** Mortality risk can be predicted in patients with ARDS by the p_aO_2/FiO_2 ratio, and lower values indicate more severe ARDS, and therefore, increased risk of mortality.

Self-assessment cases and discussion

The clinical cases described here illustrate how the measurements and calculations discussed in this chapter are used in clinical practice to guide diagnosis and management of patients with abnormal oxygenation.

Case 1: A 62-year-old woman presents to the hospital with acute onset shortness of breath and chest pain. Her initial vital signs show that she has a rapid heart rate and respiratory rate. Her O_2 sat by pulse oximetry is 88% while she is breathing room air. This improves to 96% on 2 L/min O_2 via nasal cannula and to 99% on 4 L/min O_2. Her lung exam and chest X-ray are both normal. An arterial blood gas on room air shows a pH of 7.45, pCO_2 of 32 mmHg and pO_2 of 55 mmHg.

How can the etiology of this patient's hypoxemia be determined?

It will help to review the physiologic causes of hypoxemia in Table 4.2, then narrow down the differential diagnosis based on the information provided. The patient is not at significant altitude and presents on room air, which eliminates low inspired oxygen tension. The low p_aCO_2 indicates hyperventilation, thus hypoventilation is not the cause of her hypoxemia. With a significant right-to-left shunt present, the oxygen saturation will not significantly increase with supplemental oxygen. This occurs because there is a fraction of the blood that never contacts functioning alveoli and remains hypoxemic regardless of supplemental oxygen. This patient's oxygen saturation increased from 88% to 99% on supplemental oxygen, thus no significant right-to-left shunt is present. Finally, the patient's normal chest X-ray and pulmonary exam rule out causes of significant diffusion impairment. Therefore, V/Q mismatch (high V/Q) is the likely cause of this patient's hypoxia. A V/Q scan was performed and revealed an unmatched perfusion defect consistent with a pulmonary embolism.

Case 2: A 48-year-old man presents with 5 days of fever, cough, fatigue, and shortness of breath. He is tachycardic, tachypneic, febrile, and has a room air sO_2 by pulse oximetry of 80%. This improves to 90% on 6 L/min O_2 via nasal cannula and to 94% on 100% O_2 via facemask. He has bilateral crackles over his lung fields and his chest X-ray shows diffuse bilateral infiltrates. Patient tests positive for COVID-19 and is placed on airborne isolation. Arterial blood gas on room air shows a pH of 7.30, pCO_2 of 55 mmHg, and pO_2 of 40 mmHg.

How can the etiology of this patient's hypoxemia be determined?

Again, it is helpful to think through the physiologic causes of hypoxemia and narrow down the differential diagnosis based on the information provided. The patient is not at significant altitude and is on room air, which eliminates low inspired oxygen tension as a potential cause. As the patient has an elevated pCO_2 we know there is some component of hypoventilation contributing to his hypoxemia. An alveolar-arterial oxygen (A-a) gradient is calculated to determine if hypoventilation is the only cause of this patient's hypoxemia.

A-a gradient calculation

$$\text{Alveolar } pO_2 \ (p_AO_2) = A = [p_{atm} - pH_2O_{alv}] \times 0.21 - pCO_2/0.8$$

$$\text{Alveolar } pO_2 \ (p_AO_2) = A = (760 - 47) \times 0.21 - 55/0.8 = 81$$

$$\text{Arterial } pO_2 \ (p_aO_2) = a = 40 \text{ mmHg}$$

Therefore, the A-a gradient $= 81 - 40 = 41$ (Normal: 8–20 mmHg).

The patient has an abnormal A-a gradient, which indicates that the patient's hypoxemia is not caused exclusively by hypoventilation. A component of right-to-left shunt is present since the hypoxia does not fully resolve with supplemental oxygen. The abnormal chest X-ray and physical exam suggest the presence of diffusion impairment. There may also be a component of V/Q mismatch as well, as the oxygen did partially improve with supplemental oxygen.

This patient has acute respiratory distress syndrome (ARDS) secondary to infection with COVID-19, and his case illustrates that more than one mechanism of hypoxemia may be present in a critically ill patient. Patients with ARDS have heterogeneous lung injury with areas of densely consolidated lung that receives perfusion but gets no alveolar ventilation (shunt), alveoli partially filled with fluid and debris (diffusion impairment), and microthrombi in vessels that lead to poor perfusion in ventilated alveoli (V/Q mismatch and increased dead space fraction).

Case 2 continued: The patient continues to worsen and has progressive hypoxemia despite noninvasive ventilation and high flow nasal cannula. He ultimately requires intubation and initiation of invasive mechanical ventilation. After 4 h on the ventilator, he is receiving 100% FiO$_2$ and his arterial blood gas shows a pH of 7.25, pCO$_2$ of 59 mmHg, and pO$_2$ of 50 mmHg.

How can this information be used to help guide therapy?

The p_aO$_2$/FiO$_2$ ratio defines the severity of ARDS, predicts mortality risk and need for more advanced or "rescue" therapies, such as extracorporeal membrane oxygenation (ECMO). The calculations are helpful in determining a management plan and counseling patients and families about prognosis.

p_aO$_2$/FiO$_2$ ratio calculation

$$p_a O_2/FiO_2 \text{ ratio} = 50/1.0 = 50$$

A p_aO$_2$/FiO$_2$ ratio of 50 is consistent with severe ARDS, which carries an expected risk of mortality of approximately 45%.

Measurement and calculation of oxygenation parameters are extremely important in the management of patients with hypoxemic respiratory failure, and they can change rapidly as the patient's condition either improves or deteriorates. Understanding the clinical implications of these values is imperative to appropriate and timely management of hypoxemic patients.

Case 3: A 76-year-old man is found unconscious in his home by a neighbor checking on him after an especially cold night. He is taken by ambulance to the emergency department where he is found to be hypotensive and tachycardic. His room air sO$_2$ by pulse oximetry is 100%. He is intubated for airway protection. IV fluids and vasopressors are administered and his initial labs are notable for a markedly elevated lactate. His arterial blood gas after intubation on 40% oxygen shows a pH of 7.10, pCO$_2$ of 30 mmHg, pO$_2$ of 180 mmHg, and %O$_2$Hb of 65%.

How do we use this information to guide further evaluation and management?

A large discrepancy between sO$_2$ by pulse oximetry and %O$_2$Hb should raise suspicion for an alternate form of hemoglobin, especially COHb, which cannot be detected by a standard noninvasive pulse oximeter. This case illustrates an example of a patient with tissue hypoxia (elevated lactate and evidence of end organ dysfunction with shock) in the absence of hypoxemia (normal/elevated pO$_2$). Cooximetry is performed and demonstrates a COHb level of 31%.

A patient with neurologic symptoms and evidence of tissue hypoxia with a COHb level >25% should be immediately initiated on 100% oxygen to increase release of CO and binding of oxygen to hemoglobin. Furthermore, hyperbaric oxygen therapy, where p_aO_2 levels can be increased two to three fold allowing more rapid reduction of COHb levels and improving tissue oxygenation is warranted when available.

Case 4: A 24-year-old man presents with severe sepsis secondary to methicillin resistant *Staph aureus* pneumonia and bacteremia. He is requiring invasive mechanical ventilation, veno-venous ECMO, continuous hemodialysis, and vasopressor support. On Day 1, his arterial blood gas shows a pH of 7.34, pCO_2 of 36 mmHg, and pO_2 of 45 mmHg with a %O_2Hb 79% and a hemoglobin of 11.2 g/dL. His cardiac output is measured at 8.0 LPM, and he has a normal lactate level. Because of his low oxygen saturation, he is treated with a beta-blocker medication to decrease his cardiac output in attempt to increase his sO_2. On Day 2, he has a pH of 7.32, pCO_2 of 31 mmHg, and pO_2 of 70 mmHg with a %O_2Hb 95% and a hemoglobin of 9.8 g/dL. His cardiac output is measured at 5.2 L. Blood lactate is slightly elevated.

How did these changes impact oxygen delivery to the tissues?

The patient had a low arterial oxygen saturation and hypoxemia due to severe ARDS from pneumonia and sepsis. He is on multiple forms of life support, and he likely has a high level of oxygen consumption. The addition of a medication to slow his heart rate and reduce his cardiac output was effective in increasing his sO_2, but did this change actually improve oxygen delivery to the tissues? To answer this question we must calculate the oxygen content of the arterial blood (C_aO_2) in both scenarios and also take into account the cardiac output (CO).

Oxygen content calculation

$$C_aO_2 = (1.36 \text{ mL } O_2/\text{g Hb} \times [\text{Hb}] \times \%O_2\text{Hb}) + (p_aO_2 \times 0.003 \text{ mL } O_2/\text{mmHg/dL})$$

$$\text{Day 1: } C_aO_2 = (1.36 \times 11.2 \times 0.79) + (45 \times 0.003) = 12.17 \text{ mL/dL} = 121.7 \text{ mL/L}$$

$$\text{Day 2: } C_aO_2 = (1.36 \times 9.8 \times 0.95) + (70 \times 0.003) = 12.87 \text{ mL/dL} = 128.7 \text{ mL/L}$$

Oxygen delivery calculation

$$DO_2 = CO \times C_aO_2$$

$$\text{Day 1: } DO_2 = 8.0 \times 121.7 = 973 \text{ mL/min}$$

$$\text{Day 2: } DO_2 = 5.2 \times 128.7 = 669 \text{ mL/min}$$

This case illustrates how improving the hypoxemia and low sO_2 in this patient, while making the clinician feel more comfortable with the numbers, did not actually improve oxygen delivery or tissue hypoxia. In fact, on day 2 when oxygenation appeared significantly improved, the actual oxygen delivery was only approximately 67% of day 1 and the patient had a higher lactate concentration on day 2.

References

1. Van de Louw, A.; Cracco, C.; Cerf, C.; Harf, A.; Duvaldestin, P.; Lemaire, F.; et al. Accuracy of Pulse Oximetry in the Intensive Care Unit. *Intensive Care Med.* **2001,** *27* (10), 1606−1613.

2. Meisenberg, G.; Simmons, W. H. *Principles of Medical Biochemistry,* 4th ed.; Elsevier Health Sciences, 2016.

3. Scott, M. G.; LeGrys, V. A.; Hood, J. L. Electrolytes and Blood Gases. In *Tietz Textbook of Clinical Chemistry and Molecular Diagnostics;* Burtis, C. A., Ashwood, E. R., Bruns, D. E., Eds., 5th ed.; Elsevier Saunders: St Louis, 2012; pp 807−835.

4. Martin, L. *All You Really Need to Know to Interpret Arterial Blood Gases,* 2nd ed.; Lippencott Williams & Wilkins: Philadelphia, 1999.

5. Stein, P. D.; Goldhaber, S. Z.; Henry, J. W. Alveolar-Arterial Oxygen Gradient in the Assessment of Acute Pulmonary Embolism. *Chest* **1995,** *107* (1), 139−143.

6. Williams, A. J. ABC of Oxygen: Assessing and Interpreting Arterial Blood Gases and Acid-Base Balance. *Br. Med. J.* **1998,** *317* (7167), 1213−1216.

7. The ARDS Definition Task Force. Acute Respiratory Distress Syndrome: The Berlin Definition. *J. Am. Med. Assoc.* **2012,** *307* (23), 2526−2533.

8. Bellani, G.; Laffey, J. G.; Pham, T.; Fan, E.; Brochard, L.; Esteban, A.; et al. Epidemiology, Patterns of Care, and Mortality for Patients With Acute Respiratory Distress Syndrome in Intensive Care Units in 50 Countries. *J. Am. Med. Assoc.* **2016,** *315* (8), 788−800.

9. Ortiz, R. M.; Cilley, R. E.; Bartlett, R. H. Extracorporeal Membrane Oxygenation in Pediatric Respiratory Failure. *Pediatr. Clin. North Am.* **1987,** *34* (1), 39−46.

10. Astrup, P.; Severinghaus, J. W. *The History of Blood Gases, Acids and Bases;* Munksgaard: Copenhagen, 1986.

11. *Radiometer pO_2 Reference Manual.*

Chapter 5

Calcium physiology and clinical evaluation

Introduction and history

In 1883, Sydney Ringer showed that calcium was essential for myocardial contraction *(1)*. In 1934, Franklin McLean and Baird Hastings published a report in the Journal of Biological Chemistry showing that the ionized calcium concentration was proportional to the amplitude of frog heart contraction, whereas protein-bound and citrate-bound calcium had no effect *(2)*. After many attempts to develop improved methodology, reliable ion-selective electrodes for measuring ionized calcium were ultimately developed and are now available.

The electrolytes calcium, magnesium, and phosphate have many roles in the structure and function of bone, the function of membranes, and the activation of numerous enzymes involved in genetic regulation, muscle contraction, and energy utilization. Parathyroid hormone, vitamin D, and calcitonin regulate the absorption, distribution, and excretion of these ions within the intestines, kidneys, bones, and soft tissues.

Other than diagnosing parathyroid dysfunction and hypercalcemia of malignancies, the usefulness of calcium measurements in diagnostic tests was of moderate interest through the 1970s. Over the past 40 years, a prominent role has evolved for monitoring calcium concentrations during certain surgeries and in critically ill patients.

Most calcium disorders are related to the following, for which calcium measurements are most useful:

- A deficiency of, or a defect in, parathyroid or vitamin D metabolism
- Malignancy, especially with hypercalcemia
- Kidney disease
- Magnesium deficiency
- Evaluating patients in critical illness, such as sepsis
- Evaluating neonates with unresolved hypocalcemia
- Excess administration of citrate, calcium, or saline

Blood Gases and Critical Care Testing. https://doi.org/10.1016/B978-0-323-89971-0.00009-4

In addition, understanding the regulation of calcium and other electrolytes by parathyroid hormone (PTH), vitamin D, and calcitonin is helpful in the differential diagnosis of many calcium disorders.

A most important point to make is that the correction of hypocalcemia is not always necessary or warranted. Abnormal plasma calcium concentrations are often a marker for the disease severity and resolve spontaneously with a resolution of the primary disease process. Furthermore, low ionized calcium concentrations may be protective, such that correction may actually be harmful *(3)*.

Calcium physiology

Approximately 99% of body calcium resides in the skeleton, of which a small percentage is freely exchangeable with the extracellular calcium. Although only 1% of the calcium in the body is present in the extracellular and intracellular spaces, the exchange and balance of calcium ions are critical in these spaces. The calcium concentrations in blood plasma are tightly regulated, with calcium ions playing a vital role in nerve impulse transmission, muscular contraction, blood coagulation, hormone secretion, and intercellular adhesion *(4)*.

Overall calcium balance is regulated by the concerted actions of calcium absorption in the intestine, reabsorption in the kidney, and exchange from bone. All of these processes are controlled by hormones that are part of the complex mechanism regulating both the total body and extracellular calcium.

The normal plasma concentration of total calcium is approximately 2.20–2.55 mmol/L. While all calcium in blood plasma are calcium ions, much of these ions are bound to anions, with about 40%–45% bound to proteins (mostly albumin), and about 10%–15% complexed to smaller anions, mostly bicarbonate, citrate, lactate, and phosphate. The remaining 45%–50% are free calcium ions *(5,6)*.

The concentration of calcium in the cytoplasm of cells is about 15,000–20,000 times lower than the concentrations in extracellular plasma *(7)*. This extracellular-to-intracellular gradient is maintained by the high-energy phosphates powering various ion pumps. The ionized calcium in the blood is the physiologically active portion of calcium and controls many neuromuscular and secretory processes *(8–10)*.

As an example, ionized calcium plays a vital role in cardiovascular function. The inward flow of calcium ions into cardiac cells helps control cardiac rhythm by binding to contractile proteins in myocardial cells, which initiates the contractile process. The rate of flow of calcium ions into smooth muscle cells influences the tension of arterioles that regulate blood pressure. Calcium channels in the cell membrane regulate this flow by opening when the membrane is depolarized. Magnesium ions help stabilize the calcium channels. Cardiotropic drugs such as epinephrine and isoproterenol facilitate the transport of calcium ions through these channels promoting more rapid or stronger contractions increasing cardiac work. Calcium channel blockers, such as nifedipine, verapamil, and diltiazem lower blood pressure by relaxing blood vessels and slowing the heart rate which increases blood flow and oxygen supply to the heart and reduces cardiac work *(4,11)*.

Calcium ions enter cells through calcium channels, triggering the release of calcium ions from both the sarcoplasmic reticulum and the inner cell membrane. Regulation of calcium entry through these voltage-gated channels is essential for the control of physiological processes such as secretion, synaptic transmission, and excitation—contraction coupling *(12)*. ATPases remove the excess calcium ions from inside the cell and transport calcium ions back into the sarcoplasmic reticulum for future release. In addition to these voltage-dependent calcium channels, other phosphorylation-dependent calcium channels are opened by a protein kinase activated by cyclic adenosine monophosphate (cAMP) *(12)*. Several substances affect these channels: the cardiotropic drugs epinephrine and isoproterenol facilitate the transport of calcium ions; acetylcholine and calcium channel blockers hinder the transport of calcium ions *(11)*.

Calcium ions are also important as second messengers. After stimulation at the cell surface by a specific molecule (first messenger), the inward flux of calcium ions (second messenger) initiates cellular actions, such as the production of hormones, such as insulin, aldosterone, vasopressin, and renin. This is done by keeping cytoplasmic Ca^{2+} concentration low at rest and by mobilizing Ca^{2+} in response to stimulus, which in turn activates the cellular reaction *(13)*.

Regulation in the blood

Intestinal (GI) absorption of calcium. Most GI absorption of calcium is in the duodenum, jejunum, and ileum. Absorption occurs by either paracellular pathways through "tight junction" channels between cells, or transcellular pathways through cells. The paracellular route is partly regulated by 1,25-dihydroxy vitamin D (calcitriol) that increases the permeability of these tight junctions to Ca ions. Calcitriol also stimulates the intestinal cells to increase the synthesis of calbindin, an intracellular protein that accepts Ca ions from the intestinal microvilli, then ultimately releases the Ca ions into blood *(4)*.

Renal regulation of calcium. As shown in Fig. 5.1, 60%—70% of filtered Ca ions are reabsorbed in the proximal convoluted tubule (PCT), mostly by passive mechanisms, but small amounts are reabsorbed by active mechanisms primarily regulated by PTH and calcitonin *(4)*. In the thick ascending limb of the loop (TAL), paracellular Ca ion absorption is driven by the Na—K—2Cl (NKCC2) cotransporter and the renal outer medullary K ion (ROMK) channel.

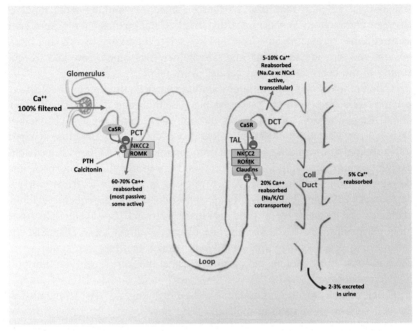

FIGURE 5.1 Processing of Ca ions in segments of the nephron 100% of Ca ions are filtered through the glomerulus, with reabsorption occurring in the proximal convoluted tubule (PCT), the thick ascending limb (TAL) of the loop, the distal convoluted tubule (DCT), and the collecting duct (Coll Duct). *CaSR*, calcium sensing receptor; *NKCC2*, Na—K—2Cl cotransporter; *ROMK*, renal outer membrane K channel. When activated, the CaSR inhibits the transporters (NKCC2 and Claudins) and ion channels (ROMK) that reabsorb Ca ions. Transcellular refers to the process where ions move through cells. PTH and calcitonin promote tubular reabsorption of Ca ions in the PCT. *Original drawing.*

Cotransporters are carrier proteins that simultaneously transport two different molecules or ions from one side of a membrane to the other at a rate of several thousand molecules per second. By using a favorable electrical or concentration gradient to transport one ion across a membrane, another ion is transported against its concentration gradient. The NKCC2 and ROMK processes transport Na, K, and Cl ions in a manner that yields a net positive charge in the lumen of the TAL, which drives the reabsorption of Ca ions by a paracellular route.

The cells in the TAL also express several "tight junction" proteins called claudins. Claudins are membrane proteins in these tight junctions between cells that regulate the flow of molecules and ions through these paracellular (between-cell) spaces. In the TAL, claudins reabsorb a small but important amount of Ca ions. Claudins are closely regulated by the calcium-sensing receptors (CaSR) that inhibit their ability to reabsorb Ca ions *(4)*.

Hormonal regulation of calcium

Three hormones regulate serum calcium by altering their secretion rate in response to changes in ionized calcium: PTH, vitamin D, and calcitonin, although the role of calcitonin is less clear. Their actions are shown in Table 5.1. A molecule with PTH-like actions, called PTH-related peptide (PTHrP), appears to have hypercalcemic actions similar to those of PTH, and may be especially important in hypercalcemia of malignancy *(14,15)*.

Parathyroid hormone. When calcium levels in blood fall, cells in the parathyroid glands respond by synthesizing and secreting PTH, which acts synergistically with vitamin D to increase blood calcium levels. The parathyroid glands respond rapidly to a decrease in ionized calcium concentration in blood, with a fourfold increase in PTH secretion stimulated by a 5% decrease in ionized calcium *(4,10,16)* and is stopped by an increase in ionized calcium *(17,18)*. PTH is synthesized in the parathyroid gland as a precursor molecule of 115 amino acids. After cleavage in the gland, intact PTH is secreted into the blood, where it circulates as an intact hormone and also as amino-terminal, carboxy-terminal, and midmolecular fragments that have varying biological activity. Both intact PTH and amino-terminal PTH interact

TABLE 5.1 Physiologic actions of PTH, vitamin D, and calcitonin.

Hormone	Actions
PTH	Promotes renal tubular reabsorption of Ca ions by inhibiting the calcium sensing receptors
	Promotes urinary excretion of PO_4
	Activates 1-a-hydroxylase in the kidney, which synthesizes active vitamin D (calcitriol). Vitamin D activates GI absorption of calcium
	Stimulates bone resorption that releases Ca ions into the ECF
Vitamin D (calcitriol)	Promotes intestinal absorption of Ca and PO_4 ions
	Promotes tubular reabsorption of Ca and PO_4 ions in the kidney
	Regulates its own homeostasis by suppressing renal production of 1-α-hydroxylase and activating the production of 25-α-hydroxylase
Calcitonin	Is produced in the thyroid gland in response to hypercalcemia
	Blocks bone osteoclasts from releasing Ca ions into the ECF
	Blocks Ca absorption in the kidney and GI tract

with PTH receptors on bone and renal cell membranes *(18)*. In bone, PTH activates osteoclasts to release cytokines that enhance the breakdown of bone to release calcium and phosphate. In kidneys, PTH conserves calcium by increasing tubular reabsorption of Ca ions and lowers phosphate by inhibiting tubular reabsorption of phosphate. PTH also activates a specific 1-α-hydroxylase enzyme in the kidney that increases the production of active vitamin D.

Vitamin D. Inactive forms of vitamin D₃ are obtained either from the diet or from exposure of skin to sunlight. These are converted in the liver to 25-hydroxycholecalciferol (25-OH-D₃), then activated by the enzyme 1-α-hydroxylase in the kidney to form 1,25-dihydroxycholecalciferol [1,25-(OH)₂-D₃; calcitriol], the biologically active form. As noted earlier, PTH promotes this conversion to active vitamin D by activating the production of the 1-α-hydroxylase enzyme, and phosphate inhibits this production. The metabolism of vitamin D is shown in Fig. 5.2.

Active vitamin D *(4)*:

- Enhances the activity of calcium pumps and calcium channels in intestinal cells to increase calcium and phosphate absorption
- Has a short-term enhancement of PTH-activated bone resorption to release calcium into the blood
- Acts with PTH to minimize renal excretion of calcium
- Contributes to the negative feedback regulation of PTH through specific receptors for 1,25-(OH)₂-D₃ on the parathyroid glands

Calcitonin. If calcium levels in the blood increase, several mechanisms act to reverse the increase: (1) C cells in the thyroid gland synthesize calcitonin, which exerts its calcium-lowering effect by inhibiting the osteoclasts in bone, which promotes deposition of calcium into bone and lowers calcium concentrations in blood. Calcitonin may also promote this effect by inhibiting the actions of both PTH and vitamin D on these cells. It is still not clear if calcitonin contributes to normal calcium homeostasis, because (a) persons without thyroid glands are still able to regulate their calcium levels in the blood, and (b) calcitonin does not change in response to small (1%–2%)

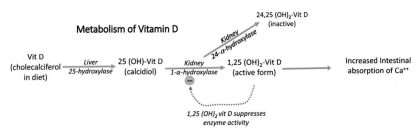

FIGURE 5.2   Synthesis and metabolism of vitamin D. The synthesis of active vitamin D (1,25 (OH)₂-vit D) begins with sequential conversion of dietary cholecalciferol by hydroxylase enzymes in the liver and kidney. As active vitamin D is formed, it suppresses the activity of 1-α-hydroxylase in the kidney. *Original drawing.*

increases in ionized calcium. A study on hemodialysis patients showed that ionized calcium had to increase by ~10% to stimulate a calcitonin response *(19,20)*. Another study noted that a 0.4 mmol/L increase was required to stimulate a response *(21)*.

Distribution in cells and blood

More than 99% of calcium in the body is in the bone as hydroxyapatite, a complex molecule of calcium and phosphate. The remaining 1% is mostly in the blood and extracellular fluid. The amount of intracellular calcium is much lower than in blood, and this is especially true of cardiac and smooth muscle cells, where the concentration of free ionized calcium in the cytosol is 10,000–20,000 times less than that of blood. Having this large gradient ensures the rapid inward flux of calcium ions necessary to trigger muscle contraction. Maintaining this gradient also requires efficient mechanisms for pumping out calcium ions. Loss of this very large gradient will lead to cell death *(7)*.

Calcium in the blood is distributed among several forms: free calcium ions, bound to protein, and bound to smaller anions *(5,6)*. As shown in Fig. 5.3, bound forms of calcium are in equilibrium with the free calcium ions. This distribution can change in many diseases, but major changes may occur during surgery or in critically ill patients because of changes in citrate, bicarbonate, lactate, and phosphate. These changes, in addition to the effect of pH on Ca^{2+} binding to proteins, are the principal reasons why ionized calcium cannot be reliably calculated from total calcium measurements in acutely ill individuals. As a rough guide, pH changes influence the concentration of ionized calcium, with a decrease of 0.1 pH unit increasing ionized calcium by ~0.05 mmol/L (0.2 mg/dL) *(22)*.

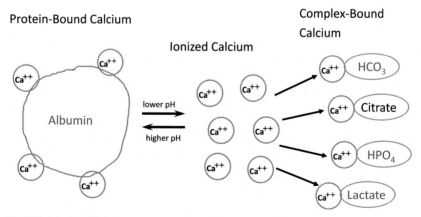

FIGURE 5.3 Equilibria between protein-bound, complex-bound, and ionized calcium. Calcium ions bind to both larger proteins, mostly albumin, and to smaller anions, mostly bicarbonate, citrate, phosphate, and lactate. *Original drawing based on Fig. 2.2 from 2nd edition.*

Hypocalcemia

Signs and symptoms of hypocalcemia

Hypocalcemia is considered to be a total calcium <8.5 mg/dL (<2.12 mmol/L) or an ionized calcium <1.00 mmol/L. As ionized calcium decreases, concentrations of 0.50–0.75 mmol/L may require administration of calcium and concentrations less than 0.50 mmol/L may produce tetany or life-threatening complications *(9)*. Ionized calcium concentrations below 0.70 mmol/L have been associated with higher morbidity and mortality *(17,23)*.

Symptoms of hypocalcemia are often manifested as cardiovascular disorders such as cardiac insufficiency, hypotension, and arrhythmias. Hypocalcemia can also cause irregular muscle spasms, called tetany, and numbness around the mouth. The rate of fall in ionized calcium initiates symptoms as much as the absolute concentration of ionized calcium *(10)*.

Causes of hypocalcemia

Prompt recognition of abnormal ionized calcium is especially important in patients with sepsis, seizures, arrhythmias, hypotension, heart failure, shock, or burns.

Hypocalcemia often occurs in patients with the following conditions *(10)*:

- Chronic or acute kidney failure
- Magnesium deficiency
- Parathyroid gland insufficiency
- Acute pancreatitis
- Critical illnesses
- During surgery
- Blood transfusion
- Hepatic disease
- Hungry bone syndrome
- Neonates

These and other frequent causes of hypocalcemia are presented in Table 5.2.

Kidney disease leading to vitamin D deficiency. Patients with kidney disease often have altered concentrations of calcium, phosphate, albumin, magnesium, and H^+ (pH). Both hyperphosphatemia and hypocalcemia are often features of renal disease, most likely due to complex alterations in the production and/or response to both PTH and vitamin D.

In chronic renal disease, secondary hyperparathyroidism often develops to compensate for hypocalcemia caused by diminished vitamin D production, due to hyperphosphatemia that inhibits synthesis of renal 1α-hydroxylase. Monitoring and controlling ionized calcium in renal disease and renal dialysis

TABLE 5.2 Frequent causes of hypocalcemia.

- Chronic renal failure
- Vitamin D deficiency or defect
- Hyperphosphatemia
- Hypomagnesemia
- Hypoparathyroidism: Postsurgical or pseudohypoparathyroidism
- Acute pancreatitis
- Patients with sepsis, burns, or other critical illness
- Following surgery, especially if large amounts of citrated blood are given
- Hepatic disease
- Hungry bone syndrome
- Illnesses in neonates
- Hypermagnesemia

may avoid problems due to hypocalcemia, such as osteodystrophy, cardiovascular instability, or problems arising from hypercalcemia such as soft tissue calcifications *(10,24)*.

Hypomagnesemia. As hypomagnesemia in hospitalized patients has become more frequent, chronic hypomagnesemia has also become recognized as a frequent cause of hypocalcemia. Hypomagnesemia may cause hypocalcemia by:

- Inhibiting the glandular secretion of PTH across the parathyroid gland membrane *(10,25)*.
- Impairing PTH action at its receptor site on bone *(10)*.
- Causing vitamin D resistance *(10,26)*.

Hypoparathyroid conditions. The most common causes of hypoparathyroidism are due to either surgical removal of excess parathyroid tissue in parathyroid surgery or parathyroid hypofunction after thyroid surgery. This condition is usually transient (lasting <5 d) unless surgery has removed too much parathyroid tissue or has interfered with parathyroid blood supply. Serum calcium is also useful for monitoring after neck surgery until a rise in the ionized calcium indicates recovery of the parathyroid gland *(27)*. Intraoperative PTH measurements are commonly done to ensure the appropriate amount of parathyroid tissue has been removed *(28)*. Primary hypoparathyroidism acquired as a genetic or autoimmune disorder is an uncommon condition.

Pseudohypoparathyroidism is characterized by end-organ resistance to PTH *(10)*. In this condition, PTH binds to receptors on the kidney and other cells, but the cellular response to PTH is impaired, leading to elevated circulating concentrations of PTH. The defect may be in the guanine nucle-otide regulatory protein Gs-alpha, which is required to activate cAMP in the normal cellular response to PTH. The combined effects of PTH resistance and hyperphosphatemia can lead to suppression of 1α-hydroxylase activity, pro-ducing a deficiency of active vitamin D and hypocalcemia *(10,29)*.

Acute pancreatitis. Many patients with acute pancreatitis develop hypo-calcemia. While the standard mechanism often cited is the precipitation of calcium ions by free fatty acids generated during acute pancreatitis, the mechanism is certainly more complex. Parathyroid dysfunction, calcitonin release, and endotoxin release may play roles as well. A consistent cause appears to be diminished secretion of PTH, which may either be low or inappropriately normal for the degree of hypocalcemia *(9,10,29)*.

Critical illness and sepsis. Critically ill patients with sepsis, thermal burns, renal failure, and/or cardiopulmonary insufficiency, frequently have symptoms attributable to hypocalcemia, such as seizures, arrhythmias, hypotension, heart failure, or shock *(3)*. These patients also tend to have abnormal acid–base regulation, diminished protein and albumin, and ionized hypocalcemia, which may directly affect mean arterial pressure *(3,30)*. Inflammatory mediators such as cytokines likely play a role in the development of these conditions.

Hypocalcemia is found in over 50% of patients in intensive care, although there are conflicting results about whether hypocalcemia should be corrected or not *(10,31)*. Some suggest that hypocalcemia is simply a marker of disease severity so that hypocalcemia will spontaneously resolve with resolution of the primary disease process *(3,10)*. Others report that if calcium is carefully administered to maintain normal ionized calcium concentrations in blood, adverse side effects are not observed *(32)*.

During surgery. In patients undergoing surgery, adequate calcium con-centrations enhance cardiac output and maintain blood pressure. Monitoring and adjusting calcium concentrations may be warranted in open-heart surgery when the heart is restarted, because normalizing ionized calcium by admin-istering calcium is a prudent action to prevent the cardiac alterations associ-ated with hypocalcemia *(8)*.

Blood transfusion. Monitoring ionized calcium is important in patients receiving large volumes of citrated blood products, which may bind and reduce serum calcium concentrations. This can occur during surgery, trauma, or other causes of massive hemorrhage and is especially notable during liver trans-plantation when liver function (the major organ for metabolizing citrate) is compromised or absent.

Hepatic disease. More severe liver diseases that impair synthetic functions can cause vitamin D deficiency from impaired 25-hydroxylation of vitamin D, decreased production of bile salts that inhibit GI absorption of vitamin D, or decreased synthesis of vitamin D-binding protein *(10)*.

Hungry bone syndrome. Hungry bone syndrome refers to hypocalcemia caused by the deposition of calcium into the bone at an accelerated rate during the healing of osteodystrophies. This syndrome is sometimes seen after surgical correction of long-standing secondary hyperparathyroidism by partial parathyroidectomy. It may also occur from osteoblastic metastases in various malignancies.

Neonatal hypocalcemia. Ionized calcium concentrations in the blood of apparently normal neonates are high at birth, rapidly decline by 10%−20% after 1−3 days, then stabilize at concentrations slightly higher than in adults after ∼1 week *(33,34)*. Hypocalcemia in neonates is defined as follows:

- Total calcium <2 mmol/L (8 mg/dL) in term infants or <1.75 mmol/L (7 mg/dL) in preterm infants.
- Ionized calcium from 0.75 to 1.10 mmol/L (3.0−4.4 mg/dL).

Signs are primarily neurologic and include irritability, muscle twitches or tremors, poor feeding, lethargy, and seizures in more severe cases *(34)*.

Early-onset hypocalcemia is common and ordinarily resolves in a few days, and asymptomatic neonates with serum calcium levels >1.75 mmol/L (7 mg/dL) or ionized calcium >0.88 mmol/L (3.5 mg/dL) rarely require treatment. If total calcium concentrations in term infants are <1.75 mmol/L (7 mg/dL), or are <1.5 mmol/L (<6 mg/dL) in preterm infants, 10% calcium gluconate (200 mg/kg) may be given by slow IV infusion over 30 min. Too-rapid infusion can cause bradycardia, so heart rate should be monitored during the infusion *(33)*.

Slightly lower concentrations of ionized calcium in healthy infants may be a normal physiological stimulus to activate parathyroid gland function and do not require treatment.

Factors that are related to the incidence and severity of hypocalcemia in neonates include *(33,34)*:

- Prematurity or infants who are small for gestational age
- Hypomagnesemia
- Maternal diabetes
- Complications during delivery
- Perinatal or birth asphyxia

A transient hypoparathyroidism may cause hypocalcemia in preterm neonates and some small-for-gestational-age neonates. In infants of mothers with diabetes or hyperparathyroidism, the mother's higher plasma ionized calcium concentrations during pregnancy can cause a hypoparathyroid condition in the child. Perinatal asphyxia may also increase serum calcitonin, which inhibits calcium release from bone *(33)*.

Laboratory evaluation of hypocalcemia

Total calcium, "corrected" total calcium, and ionized calcium

Although total calcium is often the initial laboratory test to detect hypocalcemia, ionized calcium is a more definitive parameter for confirming a calcium abnormality. Once hypocalcemia is confirmed with a low ionized calcium concentration, the sequence of laboratory tests to elucidate the cause is shown in Fig. 5.4.

While the practice of "correcting" total calcium results based on albumin and other test results continues, numerous studies have demonstrated that this correction does not improve the diagnostic accuracy of uncorrected total calcium. As emphasized many years ago *(35)*, ionized calcium cannot be accurately calculated from total calcium. As one example, it is not uncommon for a patient with a low total calcium and low albumin (who might be presumed to be normocalcemic) to have a low ionized calcium. Numerous recent reports have reconfirmed that ionized calcium cannot be accurately predicted by correcting the total calcium concentration based on albumin concentration *(36–40)*. Hamroun et al. *(41)* wryly called this practice "The unachieved quest of a perfect calcium estimation formula."

Measuring ionized calcium is no longer a highly specialized test that is available only in advanced laboratories. There are now many less-expensive and quite accurate instruments for measuring ionized calcium in the blood, with many suitable for point-of-care and resource-limited settings.

Laboratory Tests in Evaluating Hypocalcemia

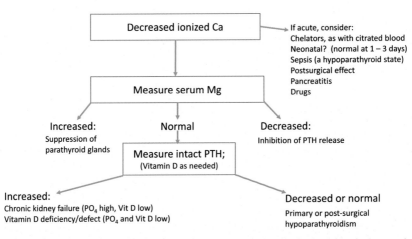

FIGURE 5.4 Laboratory tests in evaluating hypocalcemia. If the ionized calcium is decreased, the patient's history may indicate the cause, such as transfusion with citrated blood, sepsis, or following surgery. Following that, the Mg and PTH should be measured. *Original drawing.*

Treatment of hypocalcemia

As noted earlier, the correction of hypocalcemia is not always necessary or warranted. Abnormal plasma calcium concentrations are often a marker for the disease severity and resolve spontaneously with a resolution of the primary disease process. Furthermore, low ionized calcium concentrations may be protective and correction may be harmful *(3)*. The treatment also depends on the cause, the severity and rapidity of onset, and the presence of symptoms. Most cases of hypocalcemia are mild and require only supportive treatment *(10)*.

If tetany is present, calcium gluconate solution may be administered. If hypomagnesemia is also present with tetany, in addition to correcting the calcium and potassium levels, relief of tetany requires repletion of magnesium as well *(23)*. Transient hypoparathyroidism after thyroidectomy or partial parathyroidectomy may benefit from either oral or IV repletion of calcium. In chronic hypocalcemia, both oral calcium and occasionally vitamin D supplements are usually sufficient *(23)*.

Hypercalcemia

Signs and symptoms of hypercalcemia

The severity of hypercalcemia may be categorized by the total or ionized calcium concentrations, as follows *(14)*:

Mild: Total Ca: $10.5 - 11.0$ mg/dL $(2.5 - 3.0$ mmol/L); Ionized Ca: $1.4 - 2.0$ mmol/L
Moderate: Total Ca: $12 - 13.9$ mg/dL $(3.0 - 3.5$ mmol/L); Ionized Ca: $2.0 - 2.5$ mmol/L
Severe: Total Ca: $14 - 16$ mg/dL$(3.5 - 4.0$ mmol/L); Ionized Ca: $2.5 - 3.0$ mmol/L

Generally, symptoms of hypercalcemia are minimally apparent until concentrations are in the moderate and severe levels, with the severity of symptoms related not only to the absolute concentration of calcium, but also to how rapidly the increase occurred. Hypercalcemia can cause a general lethargy in patients and can affect multiple organ systems. These include the following: changes in electric conduction pathways of the heart, relaxation of GI smooth muscle causing constipation and nausea, and effects on the kidney that lead to dehydration and nephrolithiasis *(8,14,15)*. Nervous system symptoms include confusion, lethargy, diminished muscle reflexes, and possible coma with severe hypercalcemia. Kidney problems include dehydration from volume loss, with kidney stones sometimes found. GI findings include constipation and anorexia. Cardiac findings include arrhythmias and hypotension *(14)*.

Because the clinical symptoms of hypercalcemia are relatively nonspecific (confusion, muscle weakness, constipation, arrhythmia), the diagnosis must be confirmed and evaluated with laboratory measurements. A total calcium measurement >3 mmol/L (>12 mg/dL) does not usually require confirmation with an ionized calcium measurement. However, a slightly to moderately

Laboratory Tests in Evaluating Hypercalcemia

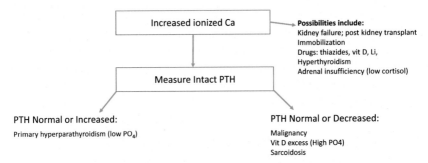

FIGURE 5.5   Laboratory tests in evaluating hypercalcemia. If the ionized calcium is elevated, the patient's history may indicate the cause, especially kidney failure, prolonged immobilization, or certain drugs. Following that, the PTH is useful in differentiating hyperparathyroidism from other causes, such as malignancy. *Original drawing.*

elevated total calcium should be confirmed by measuring ionized calcium, if available. The laboratory tests for evaluating the causes of hypercalcemia are shown in Fig. 5.5.

Causes of hypercalcemia

Malignancy and primary hyperparathyroidism (HPTH) are the most common causes of hypercalcemia, accounting for 80%−90% of all hypercalcemic patients *(42)*. Hypercalcemia due to malignancy is more likely in a hospital population, whereas primary HPTH is more common in the outpatient population.

About 20%−30% of patients with cancer have hypercalcemia at some point, which can signify an unfavorable outcome. The more common mechanisms of hypercalcemia caused by malignancies are *(14,42)*:

- Humoral hypercalcemia, sometimes caused by the production of PTHrP. This accounts for about 70%−80% of cases, which are usually more serious.
- Increased osteolytic activity with bone resorption caused by tumor activity (about 20% of cases).
- Less common is from a lymphoma secreting an active vitamin D.

Hypercalcemia related to excess PTH include:

- Primary hyperparathyroidism
- Increased release of PTH from lithium therapy.

PTHrP is a protein with some functions similar to PTH that is detected in many patients with humoral hypercalcemia of malignancy *(14,15)*. PTHrP is produced in cancers of the breast, lung, prostate, and in multiple myeloma, and has a pathological role in mediating bone destruction and hypercalcemia *(14,15,43)*. It apparently has a normal function in the fetus, activating the production of 1,25-$(OH)_2$ vitamin D in the kidney, and promoting placental transport of calcium into the fetus *(15)*.

Less common causes of hypercalcemia include the following, which may often be ruled out by clinical history:

- Prolonged immobilization for >1 week can elevate ionized calcium by 0.2−0.5 mmol/L (0.8−2.0 mg/dL) *(44)*.
- Drug therapy from thiazide diuretics, lithium, vitamin D, or antacids
- Milk alkali syndrome
- Hyperthyroidism
- Adrenal insufficiency
- Familial hypocalciuric hypercalcemia
- Sarcoidosis
- Acute renal failure

Familial hypocalciuric hypercalcemia is a genetic disorder characterized by low urinary excretion of calcium in a setting of hypercalcemia. Adrenal insufficiency causes hypercalcemia by a deficiency of cortisol. Because cortisol inhibits osteoclast activity and antagonizes the action of vitamin D, a deficiency of cortisol promotes osteoclastic activity and vitamin D action, both of which lead to hypercalcemia.

If other causes of hypercalcemia are ruled out, as shown in Fig. 5.5, the clinician must differentiate between primary HPTH and hypercalcemia due to occult malignancy. A history of chronic stable mild hypercalcemia without symptoms or weight loss favors primary hyperparathyroidism. On the other hand, an acutely rising calcium, weight loss, and fever suggest malignancy as the cause of hypercalcemia and is much more serious. Carcinomas of the bronchus, breast, head and neck, urogenital tract, and multiple myeloma account for 75% of the hypercalcemia in malignancy *(42)*.

Typical results of laboratory tests in differentiating primary hyperparathyroidism from malignancy are shown in Table 5.3. In cases of HPTH, 90%−95% of ionized calcium measurements are elevated, while total calcium is elevated in 80%−85% of cases *(45)*. For malignancies, hypercalcemia is detected much more frequently by ionized than total calcium measurements. In one study of calcium concentrations in patients with various malignancies, ionized calcium was elevated in 38% of cases, compared to only 8% for total calcium *(46)*.

In the patient with HPTH, removal of excess hyperplastic parathyroid tissue may be necessary to correct hypercalcemia. Rapid intraoperative measurements of PTH in blood have now become a standard of practice during surgical removal of excess parathyroid tissue to determine when sufficient parathyroid tissue has been removed *(28)*.

TABLE 5.3 Interpretation of laboratory tests in differentiating primary hyperparathyroidism from malignancy.

Test	Favors HPTH	Favors malignancy
Total Ca	<3.13 mmol/L (<12.5 mg/dL)	>3.13 mmol/L (>12.5 mg/dL)
Serum Cl[a]	>103 mmol/L (>103 mEq/L)	<103 mmol/L (<103 mEq/L)
Intact PTH	Elevated (or high normal)	Suppressed
Serum PO$_4$[a]	Normal to low	Variable
Hematocrit	Normal	Low
Urine Ca	High	Very high
1,25-(OH)$_2$ vitamin D	High	Low
PTHrP	Normal or undetected	Elevated

[a]*Some clinicians also use a Cl:PO$_4$ ratio as evidence for HPTH. Usually, Cl:PO$_4$ is >102 with SI units (Cl$^-$ and PO$_4$ in mmol/L) and >33 with conventional units (Cl$^-$ in mmol/L, PO$_4$ in mg/dL) in HPTH. Although often high, serum phosphate may also be low in cases with PTHrP production.*

In some patients with prolonged secondary HPTH, successful renal transplantation removes the hypocalcemic stimulation to the parathyroid gland. However, the long-hypertrophied parathyroid glands may autonomously secrete PTH, leading to tertiary HPTH. This condition resembles primary HPTH and may require surgical removal of an appropriate amount of parathyroid tissue *(47)*.

Treatment of hypercalcemia

The treatment depends on the cause and severity of hypercalcemia. Because patients who are hypercalcemic are almost always hypovolemic, volume repletion with IV crystalloid is often an essential part of care. Once volume repletion is achieved, a loop diuretic such as furosemide may be administered to increase urinary excretion of calcium. In mild hypercalcemia with total calcium <11.5 mg/dL (<2.88 mmol/L) with only mild or absent symptoms, correction of hypercalcemia may be deferred until the cause is determined. Moderate hypercalcemia with total calcium of 11.5−15 mg/dL (2.88−3.76 mmol/L) may be treated initially with saline, followed by a loop diuretic if necessary. If hypercalcemia is severe with a total Ca >15 mg/dL (3.76 mmol/L), in addition to volume expansion and diuretics, calcitonin and bisphosphonates may be required to prevent excessive calcium release from the bones. In rare cases of extremely high serum calcium >18 mg/dL (4.5 mmol/L), hemodialysis may be a necessary part of the treatment strategy *(48)*.

Interpretation of calcium and PTH measurements

In differentiating various causes of hypocalcemia and hypercalcemia, concurrent measurements of PTH and calcium are very helpful *(4,10,14)*. Fig. 5.6 shows the relationship between total calcium and intact PTH expected for primary hypo- and hyperparathyroidism, secondary hyperparathyriodism, and hypercalcemia of malignancy.

Although several assays have been developed for measuring various forms of the PTH molecule, the intact PTH assay is now used almost exclusively because it has the best combination of clinical specificity and sensitivity. It detects the entire biologically active PTH molecule of 84 amino acids, which is the major form of PTH secreted from the gland. While intact PTH has a relatively short in vivo half-life in plasma, about 5 min, it has a much longer in vitro half-life in plasma after collection and centrifugation, which enables accurate testing.

Older assays include NH_2-terminal PTH assay that detects the amino terminal (1−34 amino acid) fragment of PTH. The midmolecule and COOH-terminal PTH assays detect PTH fragments (including intact PTH) that contain the midregion or COOH-terminal end of the PTH molecule. Because many of these PTH fragments have a longer half-life in plasma (30 min), these assays are more sensitive for detecting hyperactivity of the parathyroid gland. However, PTH results with these methods may appear inappropriately high (or normal) in patients with hypoparathyroidism or hypercalcemia of malignancy.

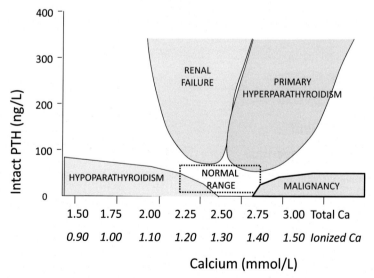

FIGURE 5.6 Relationships of calcium and PTH in various calcium disorders. This shows a plot of calcium and PTH measurements to help differentiate among potential causes of calcium disorders. *Redrawn from Fig 2.5 in 2nd edition.*

Additionally, decreased glomerular filtration also tends to spuriously elevate these PTH results by prolonging the life of these fragments in the circulation.

Proper collection and handling of samples

See Chapter 10.

Reference intervals for calcium

For total calcium, the reference interval varies slightly with age, being slightly higher in children and adolescents when bone growth is most active. In general, ionized calcium concentrations change rapidly during the first 1–3 d of neonatal life. After this, they gradually decline through adolescence until stabilizing at adult levels (see Table 5.4).

Urine reference ranges for total calcium vary with diet. Patients on an average diet should excrete 50–300 mg/d. There is no clinical significance in the measurement of ionized calcium in the urine.

TABLE 5.4 Reference intervals for calcium.

Total calcium	
Child–adolescent:	2.20–2.68 mmol/L (8.8–10.7 mg/dL)
Adult (>16 years):	2.10–2.55 mmol/L (8.4–10.2 mg/dL)
Ionized calcium	
At birth	1.30–1.60 mmol/L (5.2–6.4 mg/dL)
Neonate (<2 week):	1.20–1.48 mmol/L (4.8–5.9 mg/dL)
Child to 16 years:	1.20–1.38 mmol/L (4.8–5.5 mg/dL)
Adult (>16 years):	1.16–1.32 mmol/L (4.6–5.3 mg/dL)

Self-assessment

Self-assessment questions

1. Which clinical findings are typical of hyperparathyroidism?
   a. hyperphospatemia
   b. hypophospatemia
   c. hypocalcemia
   d. hypercalcemia
   e. b and d
2. Which of the following is NOT a cause of hypercalcemia?
   a. hyperparathyroidism
   b. malignancy
   c. vitamin D deficiency
   d. drugs
   e. endocrine disorders

3. A patient has renal failure and elevated serum phosphate and PTH levels and a chronically low serum Ca. What is the most likely diagnosis from the following choices?
   a. primary hyperparathyroidism
   b. secondary hyperparathyroidism
   c. a PTH-producing tumor
   d. tertiary hyperparathyroidism
   e. hypomagnesemia
4. The patient mentioned above has a renal transplant. Three months later, the patient's plasma calcium level is consistently elevated. What is the most likely cause?
   a. primary hyperparathyroidism
   b. secondary hyperparathyroidism
   c. a PTH-producing tumor
   d. tertiary hyperparathyroidism
   e. hypomagnesemia
5. What is the approximate normal distribution of Ca ions in blood? *CB,* complexed Ca; *PB,* protein-bound Ca.
   a. 25% PB, 10% CB, 65% ionized
   b. 10% PB, 3% CB, 87% ionized
   c. 50% PB, 3% CB, 47% ionized
   d. 40% PB, 10% CB, 50% ionized
   e. 60% PB, 5% CB, 35% ionized
6. Where is most Ca ion reabsorbed in the kidney?
   a. Collecting duct
   b. Distal tubules
   c. Proximal tubules
   d. Ascending limb of the loop
   e. Glomeruli
7. Which factor is associated with increased Ca ion reabsorption in the kidney?
   a. Ca sensing receptors (CaSRs)
   b. loop diuretics
   c. claudins
   d. decreased GFR
   e. hypercalcemia
8. How is cholecalciferol converted to active vitamin D?
   a. The thyroid gland hydroxylates at the 1 and 25 positions,
   b. The kidney hydroxylates at the 25 position, then the liver hypdroxylates at the 1 position.
   c. The kidney hydroxylates at the 1 position, then at the 25 position.
   d. The liver hydroxylates at the 1 position, then the kidney hydroxylates at the 25 position.
   e. The liver hydroxylates at the 25 position, then the kidney hydroxylates at the 1 position.

Answer Key:

1. e
2. c
3. b
4. d
5. d
6. c
7. c
8. e

Self-assessment cases

Assess the clinical status of the following patients and their laboratory results for possible disorders of calcium and/or magnesium metabolism.

Case 1. A 24-years-old woman was admitted for treatment of systemic lupus erythematosus. She complained of a weight gain of ~2.3 kg during the past week and a bloated feeling. She was started on the loop diuretic furosemide, 20 mg every 6 h. Twenty-four hours later, the patient complained of numbness and tingling in her face, hands, and feet, which rapidly progressed to acute, painful carpal spasm. Lab results at this time were as follows:

Total calcium	2.05 mmol/L (8.2 mg/dL)
Ionized calcium	1.10 mmol/L (4.4 mg/dL)
Magnesium	0.50 mmol/L (1.2 mg/dL)

The patient was treated with 6 g of $MgSO_4$ and 2 g of calcium gluconate given i.v. over 30 min. Within 60 min, the patient was free of all numbness, tingling, and spasm. Lab results 2 h later were as follows:

Total calcium	2.50 mmol/L (10.0 mg/dL)
Ionized calcium	1.30 mmol/L (5.2 mg/dL)
Magnesium	0.88 mmol/L (2.1 mg/dL)

Adapted from a case history provided by Ronald J. Elin, Chief, Clinical Pathology Department, Clinical Center, National Institutes of Health, Bethesda, MD.

Cases 2−5. Evaluate the following lab results for causes of hyper- and hypocalcemia. Consider the following possible diagnoses: hyperparathyroidism, malignancy, chronic renal disease, and postsurgery in the neck.

Lab test	Reference interval	Case 2	Case 3	Case 4	Case 5
Intact PTH (ng/L)	14–72	2	951	3	75
Total Ca					
(mmol/L)	2.10–2.55	1.88	1.98	4.10	2.65
(mg/dL)	8.41–10.2	7.5	7.9	16.4	11.8
Ionized Ca (mmol/L)	1.16–1.32	1.11	1.07	2.34	1.46
Phosphate					
(mmol/L)	0.87–1.49	1.19	2.07	1.19	0.90
(mg/dL)	2.7–4.6	3.7	6.4	3.7	2.8

Case 6. A 41-years-old female office worker with a recent diagnosis of Graves' disease (hyperthyroid condition) was admitted in January for evaluation of symptomatic hypocalcemia. She had a history of gallstones several years ago, for which she restricted her intake of fat-rich foods such as meats, oily fish, and milk. For her hyperthyroid condition, she has been on the antithyroid drug methimazole (MMZ). About 1 month ago, she felt numbness in her hands and lips, and a general fatigue. Her current lab results showed a low total Ca of 7.3 mg/dL (RI 8.4–10.4), a low PO_4 of 2.6 mg/dL (RI 2.9–4.8), a normal serum Mg, and a slightly elevated ALP of 389 U/L (RI 134–359). With this information, what would be the most likely cause of her hypocalcemia?

a. Kidney failure
b. Liver failure
c. Dietary lack of vitamin D
d. Drug induced renal loss of calcium
e. Primary hypoparathyroidism

Self-assessment and mastery discussion

Assessment of Case 1. The patient experienced an acute rapid drop in serum magnesium concentration secondary to loss from diuretic administration, which caused more severe clinical symptoms than seen with a more gradual change. The hypomagnesemia was accompanied by a mild hypocalcemia. Furosemide is a diuretic that acts primarily on the ascending loop of Henle and leads to urinary loss of magnesium. Administration i.v. of Mg and Ca relieved the symptoms of numbness and tingling.

Assessment of Case 2. These results are from a patient 24 h after parathyroid surgery for correction of hyperplasia of the parathyroid glands. Frequently, parathyroid function is impaired for several days after surgery, which results in a temporary condition of hypoparathyroidism. The decreased PTH and calcium results are typical of hypoparathyroidism. (Note that PO_4 is not always elevated in hypoparathyroidism.)

Assessment of Case 3. This is a patient with chronic renal failure who has developed secondary hyperparathyroidism with renal osteodystrophy. Note the extremely elevated intact PTH. Renal failure leads to hyperphosphatemia and abnormal vitamin D and/or PTH metabolism. Chronic hyperphosphatemia and hypocalcemia can sometimes result in extremely high concentrations of PTH, which is appropriately secreted in an attempt to increase calcium and lower phosphate. Unfortunately, this chronic condition of HPTH often causes a spectrum of bone diseases, called renal osteodystrophy.

Assessment of Case 4. These results are typical of a patient with hypercalcemia of malignancy, in which extreme hypercalcemia develops. PTHrP is likely being produced by the malignant tissue and secreted into the blood. The intact PTH result indicates that the parathyroid glands have appropriately diminished production of intact PTH.

Assessment of Case 5. This is a patient with results typical of primary HPTH: moderate elevations of calcium and PTH with slightly decreased phosphate. Although intact PTH is only slightly elevated, it should be very low with the elevated calcium concentrations. Ionized calcium is elevated slightly more (11%) than is total calcium (4%). PTH promotes renal loss of phosphate, which explains the low normal serum phosphate observed in this patient.

Assessment of Case 6. This is a case of symptomatic hypocalcemia with severe deficiency of vitamin D, most directly related to her restricted intake of fat-rich foods that would supply vitamin D. Contributing factors include lack of UV radiation during the winter along with her use of sunscreen cosmetics. The sudden onset of symptoms was likely exacerbated by the sudden correction of her Graves' disease by the MMZ drug *(Modified from reference 49)*.

References

1. Chapman, R. A.; Niedergerke, R. Effects of calcium on the contraction of the hypodynamic frog heart. *J Physiol* **1970**, *Dec;211(2):389-421*.
2. Mclean, F. C.; Hastings, A. B. A biological method for the estimation of calcium ion concentration. *Journal of Biological Chemistry* **1934**, *107*, 337−350.
3. Aberegg, S. K. Ionized Calcium in the ICU: Should It Be Measured and Corrected? *Chest* **2016**, *149* (3), 846−855.
4. Blaine, J.; Chonchol, M.; Levi, M. Renal Control of Calcium, Phosphate, and Magnesium Homeostasis. *Clin. J. Am. Soc. Nephrol.* **2015**, *10*, 1257−1272.
5. Toffaletti, J.; Gitelman, H. J.; Savory, J. Separation and Quantitation of Serum Constituents Associated with Calcium by Gel Filtration. *Clin. Chem.* **1976**, *22*, 1968−1972.
6. Toffaletti, J.; Savory, J.; Gitelman, H. J. Use of Gel Filtration to Examine the Distribution of Calcium Among Serum Proteins. *Clin. Chem.* **1977**, *23*, 2306−2310.
7. Sejersted, O. M. Calcium Controls Cardiac Function − by All Means!. *J. Physiol.* **2011**, *589* (12), 2919−2920.
8. Ariyan, C. E.; Sosa, J. A. Assessment and Management of Patients with Abnormal Calcium. *Crit. Care Med.* **2004**, *32*, S146−S154.

9. Goldberg, D. *Calcium, Ionized;* Medscape, November 2019. https://emedicine.medscape. com/article/2087469.

10. Suneja, M.; Batuman, V. *Hypocalcemia;* Medscape, August 2019. https://emedicine. medscape.com/article/241893.

11. Beckerman, J. *Heart Disease and Calcium Channel Blocker Drugs;* WebMD, 2020. https:// www.webmd.com/heart-disease/guide/heart-disease-calcium-channel-blocker-drugs#1.

12. Catterall, W. A. Regulation of Cardiac Calcium Channels in the Fight-or-Flight Response. *Curr. Mol. Pharmacol.* **2015,** *8* (1), 12−21.

13. Endo, M. Calcium Ion as a Second Messenger with Special Reference to Excitation-Contraction Coupling. *J. Pharmacol. Sci.* **2006,** *100,* 519−524.

14. Agraharkar, M. *Hypercalcemia;* Medscape, December 2020; pp 1−14. https://emedicine. medscape.com/article/240681.

15. Mundy, G. R.; Edwards, J. R. PTH-Related Peptide (PTHrP) in Hypercalcemia. *J. Am. Soc. Nephrol.* **2008,** *19,* 672−675.

16. Toffaletti, J.; Nissenson, R.; Endres, D.; et al. Influence of Continuous Infusion of Citrate on Responses of Immunoreactive PTH, Calcium and Magnesium Components, and Other Electrolytes in Normal Adults During Plateletapheresis. *J. Clin. Endocrinol. Metab.* **1985,** *60,* 874−879.

17. Ahren, B.; Bergenfelz, A. Effects of Minor Increase in Serum Calcium on the Immunoheterogeneity of PTH in Healthy Subjects and in Patients with Primary Hyperparathyroidism. *Horm. Res.* **1995,** *43,* 294−299.

18. Mundy, G. R.; Guise, T. A. Hormonal Control of Calcium Homeostasis. *Clin. Chem.* **1999,** *45,* 1347−1352.

19. Felsenfeld, A. J.; Machado, L.; Rodriguez, M. The Relationship between Serum Calcitonin and Calcium in the Hemodialysis Patient. *Am. J. Kidney Dis.* **1993,** *21,* 292−299.

20. Felsenfeld, A. J.; Levine, B. S. Calcitonin, the Forgotten Hormone: Does It Deserve to Be Forgotten? *Clin. Kidney J.* **2015,** *8,* 180−187.

21. Messa, P.; Mioni, G.; Turrin, D.; Guerra, U. P. The Calcitonin-Calcium Relation Curve and Calcitonin Secretory Parameters in Renal Patients with Variable Degrees of Renal Function. *Nephrol. Dial. Transplant.* **1995,** *10,* 2259−2265.

22. Wang, S.; McDonnell, B.; Sedor, F. S.; Toffaletti, J. G. pH Effects on Measurements of Ionized Calcium and Ionized Magnesium in Blood. *Clin. Chim. Acta* **2002,** *126,* 947−950.

23. Lewis, J. L., III *Hypocalcemia;* Merck Manuals Professional Edition, April 2020. https:// www.merckmanuals.com/professional/endocrine-and-metabolic-disorders/electrolyte-disorders/ hypocalcemia.

24. Carney, S. L.; Gillies, A. H. G. Effect of an Optimum Dialysis Fluid Calcium Concentration on Calcium Mass Transfer During Maintenance Hemodialysis. *Clin. Nephrol.* **1985,** *24,* 28−30.

25. Anast, C. S.; Winnacker, J. L.; Forte, L. F.; Burns, T. W. Impaired Release of Parathyroid Hormone in Magnesium Deficiency. *J. Clin. Endocrinol. Metab.* **1976,** *42,* 707−717.

26. Medalle, R.; Waterhouse, C.; Hahn, T. J. Vitamin D Resistance in Magnesium Deficiency. *Am. J. Clin. Nutr.* **1976,** *29,* 854−858.

27. Eleraj, D. M.; Remaley, A. T.; Simonds, W. F.; Skarulis, M. C.; Libutti, S. K.; Bartlett, D. L.; et al. Utility of Rapid Intraoperative Parathyroid Hormone Assay to Predict Severe Postoperative Hypocalcemia after Reoperation for Hyperparathyroidism. *Surgery* **2002,** *132,* 1028−1034.

28. Sokoll, L. J.; Donovan, P. I.; Udelsman, R. The National Academy of Clinical Biochemistry Laboratory Medicine Practice Guidelines for Intraoperative Parathyroid Hormone. *Point Care* **2007,** *6,* 253−260.

29. Narins, R. G., Ed. *Maxwell and Kleeman's Clinical Disorders of Fluid and Electrolyte Metabolism,* 5th ed.; McGraw-Hill: New York, 1994; p 1082.

30. Desai, T. K.; Carlson, R. W.; Thill-Baharozian, M.; Geheb, M. A. A Direct Relationship between Ionized Calcium and Arterial Pressure Among Patients in an Intensive Care Unit. *Crit. Care Med.* **1988,** *16,* 578−582.

31. Vincent, J.-L.; Jankowski, S. Why Should Ionized Calcium Be Determined in Acutely Ill Patients? *Acta Anaesthesiol. Scand.* **1995,** *39* (Suppl. 107), 281−286.

32. Zhang, Z.; Chen, K.; Ni, H. Calcium supplementation improves clinical outcome in intensive care unit patients: a propensity score matched analysis of a large clinical database MIMIC-II. *SpringerPlus* **2015,** *4,* 594. https://doi.org/10.1186/s40064-015-1387-7.

33. Dysart, K. C. *Neonatal Hypocalcemia;* Merck Manuals, December 2018. https://www. merckmanuals.com/professional/pediatrics/metabolic-electrolyte-and-toxic-disorders-in-ne-onates/neonatal-hypocalcemia.

34. *Hypocalcemia in the Newborn;* Stanford Children's Health, 2021. https://www. stanfordchildrens.org/en/topic/default?id=hypocalcemia-90-P02376.

35. Ladenson, J. M.; Lewis, J. W.; Boyd, J. C. Failure of Total Calcium Corrected for Protein, Albumin, and pH to Correctly Assess Free Calcium Status. *J. Clin. Endocrinol. Metab.* **1978,** *46,* 986−993.

36. Dickerson, R. N.; Alexander, K. H.; Minard, G.; Croce, M. A.; Brown, R. O. Accuracy of Methods to Estimate Ionized and "Corrected" Serum Calcium Concentrations in Critically Ill Multiple Trauma Patients Receiving Specialized Nutrition Support. *J. Parenter. Enteral Nutr.* **2004,** *28,* 133−141.

37. Jennichjen, S.; van der Voort, P. H. J.; Gerritsen, R. T.; Berk, J. A. M.; Bakker, A. J. Albumin-Adjusted Calcium is Not Suitable for Diagnosis of Hyper- and Hypocalcemia in the Critically Ill. *Crit. Care Med.* **2003,** *31,* 1389−1393.

38. Ridefelt, P.; Helmersson-Karlqvist, J. Albumin Adjustment of Total Calcium Does Not Improve the Estimation of Calcium Status. *Scand. J. Clin. Lab. Invest.* **2017,** *77,* 442−447.

39. Smith, J. D.; Wilson, S.; Schneider, H. G. Misclassification of Calcium Status Based on Albumin-Adjusted Calcium Studies in a Tertiary Hospital Setting. *Clin. Chem.* **2018,** *64,* 1713−1722.

40. Toffaletti, J. Predicting Ionized Hypocalcemia with Total Calcium: Can "Correction" with Logistical Modeling of Multiple Analytes Do the Trick? *J. Appl. Lab. Med.* **January 2020,** *5* (1), 1−3 https://doi.org/10.1373/jalm.2019.030197.

41. Hamroun, A.; Pekara, J.-D.; Lionet, A.; Ghulam, A.; Maboudou, P.; Mercier, A.; Brousseau, T.; et al. Ionized Calcium: Analytical Challenges and Clinical Relevance. *J. Lab. Precis. Med.* **2020,** *5,* 22−37.

42. Lafferty, F. W. Differential Diagnosis of Hypercalcemia. *J. Bone Miner. Res.* **1991,** *6* (Suppl. 2), S51−S59.

43. Cafforio, P.; Savonarola, A.; Stucci, S.; De Matteo, M.; Tucci, M.; Brunetti, A. E.; et al. PTHrP Produced by Myeloma Plasma Cells Regulates Their Survival and Pro-Osteoclast Activity for Bone Disease Progression. *J. Bone Miner. Res.* **2014,** *29* (1), 55−66.

44. Health, H., III; Earll, J. M.; Schaaf, M.; Piechocki, J. T.; Li, T.-K. Serum Ionized Calcium During Bed Rest in Fracture Patients and Normal Man. *Metabolism* **1972,** *21,* 633−640.

45. Ladenson, J. H.; Lewis, J. W.; McDonald, J. M.; Slatopolsky, E.; Boyd, J. C. Relationship of Free and Total Calcium in Hypercalcemic Conditions. *J. Clin. Endocrinol. Metab.* **1979,** *48,* 393−397.

46. Ijaz, A.; Mehmood, T.; Qureshi, A. H.; Anwar, M.; Dilawar, M.; Hussain, I.; et al. Estimation of Ionized Calcium, Total Calcium and Albumin Corrected Calcium for the Diagnosis of Hypercalcaemia of Malignancy. *J. Coll. Phys. Surg. Pak.* **2006,** *16,* 49−52.

47. Triponez, F.; Kebebew, E.; Dosseh, D.; Duh, Q. Y.; Hazzan, M.; Noel, C.; et al. Less-Than-Subtotal Parathyroidectomy Increases the Risk of Persistent/Recurrent Hyperparathyroidism after Parathyroidectomy in Tertiary Hyperparathyroidism after Renal Transplantation. *Surgery* **2006,** *140,* 997−999.

48. Lewis, J. L., III *Hypercalcemia; Merck Manuals Professional Edition,* April 2020. https://www.merckmanuals.com/professional/endocrine-and-metabolic-disorders/electrolyte-disorders/hypercalcemia.

49. K. Miyashita, T. Yasuda, H. Kaneto, et al. A case of hypocalcemia with severe vitamin D deficiency following treatment for Graves' Disease with methimazole. Case Reports in Endocrinology; volume 2013; article ID 512671: 1−4.

Chapter 6

Magnesium physiology and clinical evaluation

Introduction

In 1695, from well water in Epsom, England, Dr. Nehemiah Grew prepared Epsom salts, a name still given to magnesium sulfate. The biological significance of magnesium as a constituent of plants has been known since the 18th century, with magnesium ion an essential component of chlorophyll.

Hypomagnesemia is observed in about 10% of hospitalized patients and especially critically ill patients who often have other coexisting electrolyte abnormalities and long hospitalizations. Hypomagnesemia is associated with numerous complications including ventricular arrhythmia, coronary artery spasm, prolonged hospitalization, and increased mortality *(1)*.

While there are no practical methods to measure overall magnesium status, the measurement of total magnesium remains the usual test for assessing clinical magnesium status. Ionized magnesium methods are available, with reliability not yet to the standard of ionized calcium methods. The Mg ion sensors for these electrodes were developed from studies on over 200 ionophores produced by the late Dr. Wilhelm Simon and his colleagues.

Magnesium distribution and regulation in the blood

The human body contains ~1 mol (24 g) of magnesium, with ~53% in the skeleton and ~46% in soft tissues such as skeletal muscle, liver, and myocardium. Magnesium is primarily an intracellular ion, with only ~1% in blood and ECFs *(2)*. Like calcium, magnesium in serum exists as protein-bound (24%), complex-bound (10%), and ionized (66%) forms *(3)*. As with calcium, the pH of blood apparently affects the binding of Mg ions by proteins in the blood *(4)*.

The recommended dietary intake of magnesium is 10−15 mmol/d. Rich sources of magnesium include green vegetables, meat, grains, and seafood. The small intestine absorbs from 20% to 60% of the dietary magnesium by both active and passive transport mechanisms, depending on the need. Secretions from the upper GI tract and especially the lower GI tract contain

Blood Gases and Critical Care Testing. https://doi.org/10.1016/B978-0-323-89971-0.00006-9
111

magnesium. This explains why prolonged diarrhea, vomiting, inflammatory bowel disease, surgical removal of intestinal segments, and pancreatitis often cause magnesium depletion.

Mg homeostasis mainly involves the kidneys, the small intestine, and bones, with GI absorption and renal excretion the most important mechanisms. Active transcellular Mg uptake is through specific Transient Receptor Potential Melastatin 6 and 7 (TRPM6 and TRPM7) Mg ion channels in the small intestines, with paracellular (between cells) absorption driven by electrochemical gradients developed by ion movements such as calcium through epithelia in the small intestines *(1)*.

The kidneys are the primary site of Mg homeostasis and play a key role in regulating and maintaining Mg balance. The proximal tubule reabsorbs 15%−20% of filtered Mg, the distal tubule 5%−10%, and almost none is reabsorbed in the collecting ducts (Fig. 6.1). The thick ascending limb (TAL) of the loop is the major site, reabsorbing about 60%−70% of filtered

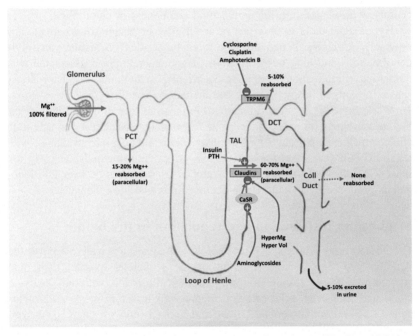

FIGURE 6.1 Processing of Mg ions in the nephron. This shows the filtration of Mg ions through the glomerulus and the reabsorption in the proximal convoluted tubule (PCT), the thick ascending limb (TAL) of the loop, the distal convoluted tubule (DCT), and the collecting duct (Coll Duct). *CaSR*, calcium sensing receptor; *Hyper Vol*, hypervolemia; *HyperMg*, hypermagnesemia; *PTH*, parathyroid hormone; *TRPM6*, Transient Receptor Potential Melastatin 6 ion channel. Paracellular refers to the process where ions move between cells. PTH and insulin promote tubular reabsorption of Mg ions; hypermagnesemia and hypervolemia inhibit tubular reabsorption. When activated, the CaSR opposes the reabsorption of Mg ions. *This is an original drawing.*

Mg *(2,5)*, promoted by several cotransporters and ion channel facilitators. These include the proteins claudin-16 and claudin-19, which promote Mg ion reabsorption in the TAL. Claudins are membrane proteins in the "tight-junctions" between cells that appropriately facilitate or block the flow of molecules and ions through the paracellular (between-cell) spaces. In the TAL, claudins reabsorb mainly Mg ions and a small amount of Ca ions. Claudins are closely regulated by the calcium-sensing receptors (CaSR) that inhibit their ability to reabsorb Mg ions. The actions of these transporters are shown in Fig. 6.1.

Magnesium absorption is fine-tuned in the distal convoluted tubule (DCT) that reabsorbs 5%−10% of filtered Mg via the Mg ion channel TRPM6. TRPM6 are protein cation channels embedded in the cells of the DCT that allow Mg ions to flow into cells. When the body needs additional Mg, the TRPM6 channels allow Mg ions to be reabsorbed from the DCT fluids into the blood. When the body has sufficient or excess Mg, the TRPM6 channels allow Mg ions to be lost from the DCT cells into the urine, and the CaSRs are activated to inhibit Mg ion absorption *(2,6)*.

The renal threshold for magnesium is ~0.60−0.85 mmol/L (~1.46−2.07 mg/dL). Because this is close to the normal serum concentration, the kidneys rapidly excrete even slight excesses of magnesium in the blood. Factors that affect tubular reabsorption of magnesium in the kidney are listed in Table 6.1.

Under normal physiologic conditions, both passive and active transport systems absorb magnesium in the GI tract, with the duodenum absorbing 11%, the jejunum 22%, the ileum 56%, and the colon 11% *(6)*. A transcellular transport mechanism relies on active transporters TRPM6 and TRPM7. These transporters strip away the hydration shell of magnesium allowing it to pass

TABLE 6.1 Factors that affect magnesium absorption in the kidney *(2)*.

Factors that increase Mg absorption	Factors that decrease Mg absorption
PTH	Hypermagnesemia
Insulin	Hypercalcemia
Calcitonin	Diuretics
Glucagon	Aminoglycoside antibiotics
ADH	Amphotericin B (antifungal)
Aldosterone	Cisplatin (chemotherapeutic)
Metabolic alkalosis	Cyclosporin (immunosuppressant)
Dietary Mg deficiency	Mutations to genes that code for Mg ion cotransporters and ion channels (claudins, TRPM6)

through the channels. Passive paracellular diffusion occurs in the small intestine and is responsible for 80%—90% of overall magnesium absorption. A high Mg concentration of 1—5 mmol/L in the lumen drives this passive transport, which relies on electrochemical and solvent diffusion to pass Mg ions through the tight junctions between intestinal cells *(7)*.

Although parathyroid hormone (PTH) has a role in regulating both magnesium and calcium, no specific mechanism for regulating Mg ion homeostasis in the blood has yet been described. Similar to its effect on calcium, PTH increases renal reabsorption of magnesium and enhances absorption of magnesium in the intestine. Paradoxically, both hypermagnesemia and hypomagnesemia can depress the secretion of PTH. Hypermagnesemia acts similarly to hypercalcemia by binding to CaSRs on the parathyroid gland and shutting off PTH production. Hypomagnesemia apparently blocks ion channels and suppresses the release of PTH, which leads to hypocalcemia. Magnesium regulation may depend on a complex interdependence of renal excretion, GI absorption, and bone exchange, and the ionized intracellular Mg concentration may be the regulatory signal *(8)*. In addition to PTH, insulin, aldosterone and vitamin D may play a role in magnesium regulation *(9,10)*.

The parathyroid gland is far more sensitive to a decrease in ionized calcium than magnesium *(3,9)*. As shown in Fig. 6.2, a 0.065 mmol/L decrease in blood ionized calcium (about 5%) increased the PTH concentration about fourfold, while a 0.03 mmol/L decrease in ultrafiltrable magnesium (also about 5%) produced no detectable PTH response, as shown in Fig. 6.3.

Regulation of magnesium is not as well characterized as that of calcium, with a number of hormones possibly having an effect on renal magnesium reabsorption. Hormones that have some effect on magnesium transport in the TAL include PTH, calcitonin, glucagon, ADH, and β-adrenergic agonists, all

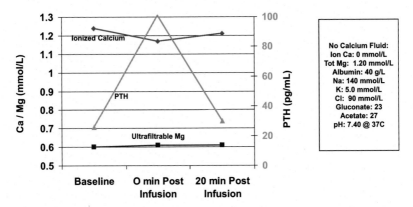

FIGURE 6.2   Changes in ionized calcium, ultrafiltrable magnesium, and PTH in healthy blood donors infused with a no-calcium fluid. The composition of the no calcium fluid is shown in the box. *This figure is from Fig. 2.6 in the 2nd edition.*

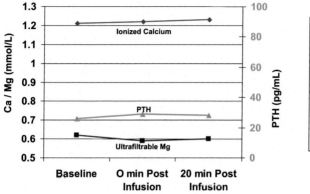

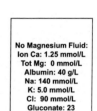

FIGURE 6.3 **Changes in ionized calcium, ultrafiltrable magnesium, and PTH in healthy blood donors infused with a no-magnesium fluid.** The composition of the no magnesium fluid is shown in the box. *This figure is from Fig. 2.7 in the 2nd edition.*

of which are coupled to adenylate cyclase in the TAL *(11)*. While the effects of these hormones may be related to an increase in luminal positive voltage and an increase in paracellular permeability, their importance in magnesium homeostasis is not known.

PTH increases the release of magnesium from bone, increases the renal reabsorption of magnesium, and, with vitamin D, enhances the absorption of magnesium in the intestine. As noted earlier, chronic or severe acute hypomagnesemia can inhibit secretion of PTH, a mechanism by which hypomagnesemia causes hypocalcemia. Aldosterone apparently inhibits the renal reabsorption of magnesium, an effect opposite that of PTH.

Insulin has a hypermagnesemia effect by increasing both intestinal and renal absorption of magnesium. Insulin has its hypermagnesemia action in the kidney as a factor that binds to receptors that activate the TRPM6 ion channels and ion transporters *(6)*.

Physiology

Magnesium is an essential cofactor in well over 300 enzymes in systems involved in almost every aspect of biochemical metabolism, such as DNA and protein synthesis, glycolysis, and oxidative phosphorylation *(1)*. Adenylate cyclase and sodium—potassium—adenosine triphosphatase (Na—K ATPase) are vital enzymes that require Mg ions for proper function. Mg also supports immune functions *(12)*, and Mg concentrations correlate with the levels of several immune mediators such as interleukin-1, tumor necrosis factor-α, interferon-γ, and the neurotransmitter peptide substance P. Moreover, Mg ions help to stabilize cell membranes and ion channels, protein and nucleic acid synthesis, regulate cardiac and smooth muscle tone, control mitochondrial functions, and support bone cell integrity *(1)*.

Magnesium ion acts as a calcium-channel blocker by affecting the influx of Ca ions at specific sites in the vascular membrane. In a healthy arterial cell with an adequate supply of magnesium, the gates are stabilized and control the entry of Ca ions. Magnesium deficiency promotes the accumulation of intracellular Ca and Na ions, leading to a state of greater contractility *(11)*.

Other conditions associated with hypomagnesemia include the following:

- The effects on myocardial function and blood pressure *(13)*.
- The implications of magnesium depletion during open-heart surgery *(14)* and critical care *(15)*.
- A possible role in migraine headaches, asthma, and chronic fatigue syndrome *(11)*.

Hypomagnesemia

Causes and clinical conditions associated with hypomagnesemia

Hypomagnesemia, defined as a serum magnesium of <0.75 mmol/L (1.8 mg/dL) *(16)*, with the common causes of hypomagnesemia shown in Table 6.2.

Critical illness

Magnesium deficiency is found in a large percentage of critically ill patients *(17)*, and the presence of hypomagnesemia on admission is associated with an increased mortality rate *(18)*. The causes of hypomagnesemia in critically ill patients are mainly from gastrointestinal disorders or renal loss of Mg. GI disorders such as prolonged diarrhea, vomiting, inflammatory bowel disease,

---

TABLE 6.2 Causes of hypomagnesemia.

- Drugs, such as cyclosporin, cisplatin, diuretics, amphotericin, aminoglycosides, proton pump inhibitors (PPIs), metformin
- Diabetes
- Dietary deficiency
- Alcoholism
- GI loss: diarrhea, malabsorption syndromes
- Cellular hypoxia
- Toxemia or eclampsia of pregnancy
- Loss through skin (burns)
- PTH deficit (sepsis or hypoparathyroidism)

surgical removal of intestinal segments, and pancreatitis are associated with magnesium depletion *(1)*. Many of the causes of Mg deficiency in ICU patients are among those listed in Table 6.2. Renal loss may be enhanced by drugs and alcoholism *(8,19)*. Although magnesium deficiency is more common in critically ill patients, hypermagnesemia is more often associated with a poor outcome than is hypomagnesemia *(20)*, most likely because hypermagnesemia is almost always a secondary effect of kidney disease that causes higher mortality.

In pediatric patients, magnesium was the most common electrolyte abnormality found in pediatric intensive care unit patients *(21)*, and ionized magnesium may be decreased in many critically ill pediatric patients who have normal total magnesium concentrations *(22)*. In this study, the use of albumin-corrected total magnesium was not reliable in estimating the ionized magnesium.

Some studies conclude that either total or ionized magnesium measurements may be used to follow magnesium status in patients in critical care *(23,24)*, while others concluded that the ionized magnesium was more specific *(20,25)*. Another study concluded that red blood cell magnesium was the best parameter to measure because it gave a higher incidence of hypomagnesemia (37%) than did either ionized magnesium (22%) or total magnesium (16%) *(26)*. See the section "Ionized Mg versus total Mg" at the end of this chapter.

In patients during abdominal surgery who were not given massive transfusions, changes in ionized and total serum Mg concentrations correlated closely *(23)*, and total serum Mg adequately screened for hypomagnesemia. Magnesium supplementation during cardiopulmonary bypass appeared to significantly benefit patients with hypomagnesemia by preventing ventricular tachycardia *(8,15)*.

In a review of 253 patients admitted to the ED, mild, moderate, and severe hypomagnesemia were found in 19.5%, 9.1%, and 2.5% of the patients, although these levels did not relate to mortality *(27)*. These were related to drugs, alcoholism, kidney failure, and poor diet.

Cardiac disorders

In cardiovascular disease, myocardial hypoxia accompanied by cellular magnesium deficit will more rapidly deplete ATP, leading to disruption of mitochondrial function and structure. With Mg ion important in energy metabolism, ion movements via Na–K-ATPase, calcium channel regulation, myocardial contraction, and other cardiac functions, the heart is particularly vulnerable to magnesium deficiency. Hypomagnesemia can disrupt mitochondrial function and production of ATP, which promote loss of myocardial potassium. These conditions can contribute to coronary vasospasm, arrhythmias, atrial fibrillation, infarction, and sudden death *(9)*. A low magnesium concentration enhances the potency of vasoconstrictive agents that can produce sustained constriction of arterioles and venules. Furthermore, a low serum magnesium promotes endothelial cell dysfunction that inhibits nitric oxide (NO) release and promotes a proinflammatory, prothrombotic, and proatherogenic environment *(14,28)*.

Table 6.3 summarizes the cardiac disorders known to be associated with loss of myocardial and/or serum magnesium *(1,13)*.

Magnesium affects vascular tone by modulating the vasoconstrictive effects of hormones such as norepinephrine and angiotensin II: A high Mg:Ca concentration ratio antagonizes their effects, whereas a low Mg:Ca ratio enhances their activity *(29)*. Ca and Mg ions compete for binding to the contractile proteins, with Ca ions initiating contraction (vasoconstriction) and Mg ions inhibiting contraction (vasodilation).

Postoperative hypomagnesemia is common following cardiac operations and is often associated with atrial fibrillation and arrhythmias *(1)*. Hypomagnesemia also impairs the release of NO from coronary endothelium, which promotes vasoconstriction and coronary thrombosis in the early postoperative period *(14)*. Intravenous magnesium sulfate solution has been shown to reduce the incidence of postoperative atrial fibrillation *(30)*. However, overdosing magnesium must be avoided *(14,31)*. In patients undergoing surgery with cardiopulmonary bypass, plasma ionized (but not total) magnesium was decreased significantly by 24 h after bypass *(32)*. By correcting serum magnesium levels during cardiopulmonary bypass, the incidence of ventricular tachycardia was reduced from 30% to 7% *(15)*.

The benefit of administering magnesium for myocardial infarction is controversial. Magnesium therapy reduced the frequency of ventricular arrhythmias in acute myocardial infarctions (AMI) *(33)*. The LIMIT-2 Study on

TABLE 6.3 Myocardial ischemic syndromes and cardiac disorders associated with loss of myocardial or serum magnesium *(1,13)*.

- Cardiac arrhythmias
- Ventricular tachycardia
- Atrial fibrillation
- Congestive heart failure
- Coronary vasospasm
- Unstable angina
- Myocardial infarction
- Alcoholic cardiomyopathy
- Sudden death ischemic heart disease

over 2000 patients found a distinct benefit of administering Mg early to patients with suspected AMI *(34)*, and a rationale for the benefit of magnesium was presented *(35)*. A key factor was thought to be the timing of magnesium supplementation, which might be required to achieve benefits in high-risk patients, before reperfusion occurs. However, the very large MAGIC trial did not find a benefit of early administration of magnesium *(36)*. A more recent review was inconclusive about magnesium being routinely administered for AMI *(37)*.

In a study of pediatric patients undergoing surgery for congenital heart defects, magnesium supplementation was so effective in preventing ectopic tachycardia (none in 13 patients) compared to the placebo group (4 in 15), that the study was terminated after 28 patients *(38)*.

Drugs

Several drugs, including diuretics, gentamicin and other aminoglycoside antibiotics, cisplatin, and cyclosporine increase the renal loss of magnesium and frequently result in hypomagnesemia. Loop diuretics (furosemide, and others) significantly increase magnesium excretion by inhibiting the electrical gradient necessary for magnesium reabsorption in the TAL. Long-term thiazide diuretic therapy also may cause magnesium deficiency, through enhanced magnesium excretion by reducing levels of the TRPM6 ion channels in the TAL cell membranes *(8,9,11)*.

Many nephrotoxic drugs, including cisplatin, amphotericin B, cyclosporine, and tacrolimus, can cause profound hypomagnesemia by increasing urinary magnesium wasting, possibly involving a variety of mechanisms that inhibit the TRPM6 channels. Good urine flow should be maintained during therapy with these drugs to prevent kidney toxicity. The effects of these and other drugs that cause hypomagnesemia are presented in a review *(39)*. Urinary magnesium wasting caused by the administration of calcineurin inhibitors such as cyclosporine and tacrolimus is partly the reason that hypomagnesemia frequently develops after kidney transplantation. Other causal factors in these patients include posttransplantation volume expansion, metabolic acidosis, and insulin resistance *(40)*.

Aminoglycosides such as gentamicin inhibit reabsorption of magnesium in the renal tubule. Because hypomagnesemia intensifies the toxic side effects of these drugs, hypomagnesemia should be avoided. Aminoglycosides are thought to produce magnesium wasting by inducing the activity of the CaSRs on the ascending limb and distal tubule of the kidney *(11)*.

Proton pump inhibitors (PPI) such as omeprazole (Prilosec) and lansoprazole (Prevacid) are often used to treat ulcers and gastroesophageal reflux disease (GERD) by reducing stomach acid production. However, with long-term use, they can cause hypomagnesemia by inhibiting the Mg ion transporters TRPM6 and TRPM7 in the intestines *(1,2)*.

Diabetes

Hypomagnesemia is common in patients with diabetes *(9,41)*, with a serum Mg <0.7 mmol/L strongly associated with type 2 diabetes. Patients with type 2 diabetes and hypomagnesemia show a more rapid disease progression and development of insulin resistance. Consequently, patients with type 2 diabetes and hypomagnesemia enter a vicious circle in which hypomagnesemia causes insulin resistance and insulin resistance causes hypomagnesemia *(6)*. Hypomagnesemia may also intensify complications frequently associated with diabetes, such as retinopathy, hypertension, cardiovascular disease, and increased platelet activity and thrombosis. In the pediatric population, hypomagnesemia was a strong, independent risk factor for insulin resistance in obese children *(42)*. Dietary Mg supplementation for many patients with type 2 diabetes improved glucose metabolism and insulin sensitivity *(10)*.

In a study of 128 patients with chronic renal failure, those with diabetes had both total and ionized Mg about 10% lower than in the nondiabetic renal patients *(43)*, and another report concluded that a low serum Mg was associated with a more rapid decline of renal function in patients with type 2 diabetes *(44)*.

The mechanism of magnesium deficiency in diabetes appears to be abnormal intracellular—extracellular distributions of Mg ions caused by insulin and other factors. Intracellular Mg ions regulate glucokinase, potassium-ATP (KATP) channels, and L-type (long-lasting voltage-dependent) Ca ion channels in the pancreatic β-cells, which activate insulin secretion. Such mechanisms make intracellular Mg ions a direct factor in the development of insulin resistance.

Insulin helps regulate Mg ion homeostasis in the kidney, where it activates both renal Mg ion channels and transporters (such as TRPM6) that determine the final urinary Mg excretion in the DCT *(6)*. Magnesium deficiency may also be related to increased urinary loss of magnesium from osmotic diuresis caused either by glycosuria or hormonal imbalances, such as decreased PTH and altered vitamin D metabolism.

In 1992, the American Diabetes Association issued a statement about magnesium and diabetes *(45)*. At that time, the ADA stated that there was a strong relationship between hypomagnesemia and insulin resistance and that magnesium supplements should be given to all patients with documented hypomagnesemia and symptoms, but did not recommend magnesium supplementation to all patients with diabetes. In 2013, the ADA issued a statement emphasizing the importance of a balanced diet to promote healthful eating patterns, emphasizing a variety of nutrient-dense foods in appropriate portion sizes, to improve overall health and specifically to attain individualized glycemic, blood pressure, and lipid goals. Their brief statement about magnesium supplementation was weakly put: "Evidence from clinical studies evaluating magnesium … supplementation to improve glycemic control in people with diabetes is conflicting" *(46)*.

Dietary deficiency

Dietary surveys show that almost half of people in the United States consume less than recommended amounts of magnesium, with older men and adolescent males and females most likely to have low intakes. Green vegetables such as spinach contain magnesium in the chlorophyll of these plants. Beans and peas, nuts and seeds, and whole, unrefined grains are also good sources of magnesium *(11)*. Adequate dietary Mg is continually necessary because the GI tract cannot increase Mg absorption during dietary Mg shortage.

Alcoholism

Chronic alcoholism has long been associated with hypomagnesemia *(47,48)*. Both acute and chronic alcohol consumption increase renal magnesium excretion. Hypomagnesemia in alcoholic patients apparently results from alcohol-induced tubular dysfunction along with a combination of dietary magnesium deficiency, vomiting, and diarrhea. Infusion of glucose solutions induces insulin secretion, which shifts magnesium back into cells. Both total and ionized magnesium apparently increase toward normal after 3 week of abstinence from alcohol *(48)*.

Cellular hypoxia

Cellular hypoxia leads to depletion of ATP, which causes cellular loss of Mg and K ions. Chronic hypoxic conditions, such as decreased cardiac output, can also result in hypomagnesemia.

Preeclampsia of pregnancy

Preeclampsia is a hypertensive disorder that occurs in ∼5% of pregnancies and causes significant morbidity and mortality. The pathogenesis of preeclampsia is complex, but it begins with abnormal placental implantation and development that leads to endothelial and immunologic dysfunction. Clinical features include the following: hypertension, proteinuria, and dysfunction of the kidneys, heart, lungs, and CNS. Also, fetal growth may be affected.

Because magnesium requirements increase during pregnancy, hypomagnesemia is common in preeclampsia and eclampsia (development of seizures). Treatment with magnesium salts is a long-accepted practice for these conditions, although extreme hypermagnesemia sometimes results from excessive doses *(49)*. The fetus may also be affected by hypermagnesemia in such cases.

Diarrhea

As noted in the section "Critical Illness," secretions from the lower GI tract are relatively rich in Mg so that diarrhea, malabsorption syndromes, bowel resection, etc. are common causes of Mg depletion.

Burns

Loss of skin integrity via burns is associated with a general loss of fluid and electrolytes.

PTH deficiency

Because PTH increases tubular reabsorption of Mg in the kidney, a deficit of PTH, such as in primary hypoparathyroidism, will lead to hypomagnesemia. Sepsis inhibits PTH secretion and is also associated with hypomagnesemia *(50)*.

Other diseases

Magnesium deficiency may promote the development and progression of renal stone formation and other renal calcification *(51)*. Magnesium ions may prevent calcification by chelating anions that would otherwise form insoluble salts with calcium.

Biochemical manifestations of hypomagnesemia

Hypokalemia. Hypokalemia is common in patients with Mg deficiency and vice versa. Indeed, patients with Mg depletion have a renal loss of potassium caused by an increased potassium secretion in the connecting tubule and the cortical collecting tubule. In the kidneys, K^+ is absorbed across the basolateral membrane via Na–K ATPase and secreted into the lumen of the connecting tubule and cortical collecting tubule. This process is mediated by renal outer-membrane potassium channels (ROMK), which require ATP. At normal intracellular Mg ion concentrations, these channels move more K ions into cells. With a lack of intracellular Mg ions, K ions move freely through the ROMK channels, increasing cellular potassium efflux that may be detected as increased potassium in the urine *(52)*.

Hypocalcemia. Mg deficiency is often associated with hypocalcemia *(1)*. Patients with combined hypocalcemia and hypomagnesemia often have low levels of PTH. This is caused by Mg deficiency inhibiting the release of PTH from the parathyroid gland. Because parenteral Mg stimulates PTH secretion, reduced PTH secretion contributes to hypocalcemia in Mg deficiency *(53)*.

Evaluation of magnesium status in patients

Magnesium should be measured during the initial examination of all patients with poor food intake, malabsorption disorders, hypokalemia, or hypocalcemia, as well as those taking magnesium-depleting agents, such as diuretics, alcohol, or aminoglycosides. A patient with hypomagnesemia may show a variety of nonspecific symptoms, such as weakness, muscle cramping, and rapid heartbeat *(54)*. While measurements of total magnesium concentrations in serum remain the usual diagnostic test for detection of magnesium abnormalities, it has limitations:

- 25%−30% of magnesium is protein-bound. Therefore, as with total and ionized calcium, total magnesium may not reflect the physiologically active free ionized magnesium. This is discussed later in the section "Ionized magnesium versus total magnesium."
- Because Mg is primarily an intracellular ion, serum concentrations will not necessarily reflect the intracellular status of magnesium *(7)*. Even when tissue and cellular magnesium is depleted by as much as 20%, serum magnesium concentrations may remain normal.

The magnesium load test may be the definitive method for detecting body depletion of magnesium, as was shown in 13 patients with pancreatitis and a normal concentration of magnesium in serum *(55)*. However, because a magnesium load test requires >48 h to complete, its clinical use is limited. After collection of a baseline 24-h urine, 30 mmol (729 mg) of magnesium in a 5% dextrose solution is administered intravenously (i.v.) as another 24-h urine is collected. Individuals with adequate body stores of magnesium will excrete 60%−80% of the magnesium load within 24 h, while magnesium-deficient patients will excrete <50% *(1)*.

Urinary magnesium may help confirm the cause of a magnesium deficit. If it is >25−30 mg/d, renal loss is suspected; if <20 mg/d, nonrenal causes should be suspected, such as inadequate intake or GI loss.

In addition to renal loss, acute hypomagnesemia can result from intracellular shifts of magnesium after the administration of glucose or amino acids *(54)*. This effect is pronounced after starvation or insulin treatment for hyperglycemia.

Treatment of hypomagnesemia

For patients at risk for magnesium depletion, magnesium supplements may be given to prevent hypomagnesemia. Magnesium replacement therapy may be warranted if serum Mg is <0.5 mmol/L (<1.2 mg/dL) *(53)*. However, because serum Mg does not necessarily reflects the total body Mg status, patients at risk for magnesium deficiency or with symptoms consistent with hypomagnesaemia should be considered for treatment even with normal serum Mg concentrations *(56)*. In acute-care patients, hypomagnesemia is often associated with coronary artery disease and coronary bypass surgery, malignancy, chronic obstructive pulmonary disease, and alcoholism. Among chronic diseases, alcoholism, liver disease, and carcinoma were commonly associated with hypomagnesemia *(19)*.

If magnesium infusion is necessary, as a guide, 25 mmol of $MgSO_4$ solution given intravenously over 12 h increased serum magnesium concentration to about 1.5 mmol/L, while 50 mmol per 12 h increased serum magnesium to about 2 mmol/L *(57)*. Renal function should be assessed to guide supplementation and prevent hypermagnesemia *(57)*. In another study, administration of 1 g of $MgSO_4$ in solution to critically ill patients increased total Mg by 0.11 mmol/L and ionized Mg by 0.05 mmol/L *(58)*.

In general, mild hypomagnesemia with no or only mild symptoms can be treated with an oral supplement, whereas parenteral Mg supplementation is indicated if Mg concentration is <0.5 mmol/L or if the patient presents with significant symptoms *(59)*. For critically ill patients with mild-to-moderate hypomagnesemia, empirically administering 1 g (4 mmol) of IV Mg will increase the serum Mg concentration by 0.08 mmol/L within 18–30 h. Trials of patients with acute myocardial infarction suggest an initial bolus (e.g., 2 g (8 mmol)) followed by continuous infusions of up to 16 g (65 mmol) over 24 h. Guidelines for treating severe hypomagnesemia are that lower doses (<6 g $MgSO_4$) can be given over a period of 8–12 h, whereas higher doses may be administrated over a period of 25 h or longer *(60)*. These timings are made especially important due to the combined slow distribution of Mg in tissues and the rapid renal excretion of magnesium *(1)*.

When drugs that deplete magnesium, such as gentamicin, diuretics, cisplatin and cyclosporine are given, $MgSO_4$ solution can be given intravenously once or more daily, as guided by serum magnesium measurements. Women with toxemia of pregnancy (eclampsia) are sometimes given very large doses of magnesium sulfate intravenously (175 mmol/24 h). This should be monitored with measurements of serum magnesium, because potentially dangerous hypermagnesemia to concentrations of 4.0 mmol/L and above can occur *(57)*.

Hypermagnesemia

Causes of hypermagnesemia

Significant hypermagnesemia is almost always associated with impaired kidney function, because healthy kidneys can readily excrete excess magnesium loads. A well-functioning kidney handles a magnesium load by diminishing Mg absorption in the thick ascending limb of the loop, where over 50% of filtered magnesium is normally reabsorbed *(49)*. While mild elevations in plasma magnesium may be seen in 10%–15% of hospitalized patients, such patients usually have some degree of impaired kidney function. More severe hypermagnesemia can occur when excessive doses of magnesium-containing antacids, enemas, or parenteral nutrition are administered to patients with renal insufficiency *(62)*. Hypermagnesemia is sometimes seen in adrenal insufficiency, probably related to diminished aldosterone causing volume depletion and hemoconcentration. The most common cause of dramatically increased plasma magnesium without chronic renal failure is found in treatment for eclampsia of pregnancy, for which large amounts of magnesium salts are given.

Kidney impairment. As kidney function declines, plasma magnesium levels rise because the only regulatory system that is able to remove magnesium is urinary excretion *(49,61)*. In patients with end-stage kidney disease or on dialysis, magnesium intake must be carefully monitored to maintain control of the plasma magnesium concentration. For example, hypermagnesemia (defined as a serum magnesium greater than 1.5 mmol/L) can occur with magnesium intake as low as 280 mg/day, which is within the range of average intake by the general population of about 200−340 mg/day, depending on weight. Because symptomatic hypermagnesemia can be caused by magnesium given as antacids or laxatives in typical therapeutic doses *(49)*, these agents should be avoided in patients with renal impairment.

Magnesium infusion. In pregnant women with severe preeclampsia or eclampsia, parenteral magnesium is commonly given intravenously to decrease neuromuscular excitability *(49)*. The usual plasma concentration achieved is 6−8.4 mg/dL or 2.5−3.5 mmol/L, but much higher levels can occur. Frequent complications of hypermagnesemia are hypocalcemia and hyperkalemia, as hypermagnesemia can suppress the release of PTH *(63)*. Neonates with hypermagnesemia may also have hyperkalemia and hypocalcemia, along with hypotonia, all of which increase the need for neonatal intensive care *(64)*.

Oral ingestion. Ingestion of very large oral doses of magnesium may exceed renal excretory capacity, especially if either chronic kidney disease or acute kidney injury is present *(65)*. Accidental overdose can also occur with over-the-counter products, such as Epsom salts ($MgSO_4$) or laxatives *(66)*.

Enemas. Enemas usually contain high concentrations of $MgSO_4$, so that substantial quantities of magnesium can be absorbed from the large bowel following an enema. Even in normal subjects, enemas containing 400−800 mmol of $MgSO_4$ can raise the plasma magnesium concentration to 7.2−19.2 mg/dL (3−8 mmol/L). This amount of magnesium given as an enema to patients with kidney failure can be fatal *(67)*.

Symptoms of hypermagnesemia

Most symptoms of hypermagnesemia are from neuromuscular effects, cardiovascular effects, and hypocalcemic effects. Symptomatic hypermagnesemia may occur with the following plasma concentrations of magnesium *(49)*:

- 2−3 mmol/L (4.8−7.2 mg/dL): Nausea, flushing, headache, lethargy, drowsiness, and diminished deep tendon reflexes.
- 3−5 mmol/L (7.2−12 mg/dL): Somnolence, hypocalcemia, absent deep tendon reflexes, hypotension, bradycardia, and electrocardiogram (ECG) changes.
- Above 5 mmol/L (12 mg/dL): Muscle paralysis leading to quadriplegia, apnea, and respiratory failure that usually precedes complete heart block and cardiac arrest.

Neuromuscular effects are the most common complication of hypermagnesemia, because increased magnesium decreases impulse transmission across the neuromuscular junction, and this can be detected by diminished deep tendon reflexes, which are usually noted when the plasma magnesium concentration reaches 2–3 mmol/L (4.8–7.2 mg/dL) *(49)*.

Cardiovascular effects are related to excess intracellular Mg blocking cardiac K channels. Hypotension, conduction defects, and bradycardia begin to appear at a plasma magnesium concentration of 2–2.5 mmol/L (4.8–6 mg/dL). ECG changes may occur at concentrations of 2.5–5 mmol/L (6–12 mg/dL). Complete heart block and cardiac arrest may occur at a plasma magnesium concentration above 7.5 mmol/L (18 mg/dL).

Moderate hypermagnesemia can inhibit the secretion of PTH, leading to a decreased plasma calcium concentration. The fall in the plasma calcium concentration is usually transient and asymptomatic, although ECG abnormalities may occasionally occur.

Treatment of hypermagnesemia

The approach to therapy largely depends on kidney function. If kidney function is normal, stopping the source of magnesium will promptly restore normal plasma magnesium concentrations. If necessary, loop or thiazide diuretics can be used to increase renal excretion of magnesium. In patients with mild-to-moderate degrees of either chronic or acute kidney impairment (eGFR between 15 and 45 mL/min/1.73 m^2), urinary elimination of magnesium may be limited. Again, the initial treatment begins with removing magnesium-containing medications followed by administering intravenous isotonic fluids such as saline plus a loop diuretic. In patients with end-stage kidney failure (eGFR less than 15 mL/min/1.73 m^2, dialysis may be the only option *(49)*.

Proper collection and handling of samples

See Chapter 10.

Diagnosis of magnesium abnormalities

While measurement of total magnesium concentrations in serum remains the usual diagnostic test for the detection of magnesium abnormalities, total magnesium concentrations have two limitations. First, about 30% of magnesium is protein bound. Therefore, as with total versus ionized calcium, total magnesium may not reflect the physiologically active ionized magnesium. Second, and probably of greater significance, because magnesium is primarily an intracellular ion, concentrations of either total or ionized magnesium in blood will not necessarily reflect the intracellular and overall body stores of magnesium. Symptomatic hypomagnesemia may not appear until serum total magnesium levels fall below 0.5 mmol/L (well below the typical reference limit), which make associations between serum magnesium concentrations and the benefit of treatment difficult *(8)*.

The *magnesium retention test* assesses overall magnesium status, but its use may be restricted to special circumstances when there is a strong suspicion of magnesium deficiency despite a normal serum magnesium. A magnesium retention test requires over 48 h to complete, collection of two 24-h urines, and giving 30 mmol of $MgSO_4$ intravenously over 12 h. If more than 60%−70% of Mg is excreted in the urine following an intravenous load, Mg deficiency is unlikely, while magnesium-deficient patients excrete less than 50% of the administered magnesium *(1,8)*. Normal renal handling of Mg is necessary for this test to be useful because increased magnesium losses due to diabetes, medications or alcohol ingestion may result in a false-negative test. If kidney function is compromised, one may see false-positive results, so the Mg retention test should not be used in patients with renal impairment or in transplant patients receiving cyclosporine or tacrolimus, both of which cause urinary Mg wasting *(68)*.

Ionized Mg versus total Mg

The routine measurement of ionized magnesium has been available for many years since sensors for magnesium ion electrodes were developed from lengthy studies on over 200 ionophores produced by the late Dr. Wilhelm Simon and his colleagues *(69)*. However, ionized magnesium measurements have more analytical variability than ionized calcium measurements because all Mg ion-sensitive electrodes are also sensitive to Ca ions. While this interference by Ca ion is corrected by simultaneous measurement of the Ca ion concentration, this additional test adds variability to the ionized magnesium result *(70)*.

There are also a few studies relating intracellular levels of magnesium to clinical information. For example, one study concluded that red blood cell magnesium was the best parameter to measure because it gave a higher incidence of hypomagnesemia (37%) than did either ionized magnesium (22%) or total magnesium (16%) *(26)*. However, regardless of its possible clinical value, at least for the foreseeable future, these measurements are not practical for clinical use.

While some studies conclude that measurements of ionized magnesium improve clinical accuracy, others find that either total or ionized magnesium provides about the same clinical value in detecting hypomagnesemia. Conclusions are often based on whether the total serum magnesium and ionized Mg do or do not correlate well in various disease states.

While it is logical to equate the physiologic activity of ionized Mg with better clinical reliability, there are other issues to consider. Total magnesium measurements have better analytical reliability and have far greater familiarity in clinical practice. However, because ionized magnesium measurements are on "blood gas" analyzers, they have the advantage of faster turnaround times and can be used in point-of-care settings. Hopefully, ionized magnesium will become more widely used and gain clinical familiarity.

Here are representative reports that conclude the total and ionized Mg results give equivalent clinical information.

Saha et al. reported strong correlations between total and ionized Mg concentrations in serum from hemodialysis patients, and in patients with intestinal disease, alcoholic liver disease, and chronic renal disease *(71,72)*.

Lanzinger et al. often found that, in patients during abdominal surgery who did not have massive transfusions, the changes in ionized and total serum Mg concentrations correlated closely *(23)*. Even though total serum Mg slightly overestimated the prevalence of hypomagnesemia, it adequately screened for hypomagnesemia.

Koch et al. found a strong correlation between ionized and total Mg ($r = 0.90$), with hypomagnesemia in 21% of ICU patients measured by ionized Mg and 18% of ICU patients measured by total Mg. They concluded that ionized Mg could be inferred from total Mg *(24)*.

In a study on 286 patients of indigenous and nonindigenous Australian descent, with or without diabetes, Longstreet and Vick noted significant correlations between total and ionized magnesium measurements in all groups ($r = 0.75$; $P < .001$). Among people with diabetes, the correlation coefficient (r) was 0.81 whereas in nondiabetics, the r was 0.66. Based on the strong correlation between total and ionized Mg concentrations in serum, they concluded that either gives an accurate assessment of magnesium status in health and chronic diabetes, irrespective of ethnicity *(73)*.

Here are representative reports that conclude that ionized Mg measurements provide better clinical information.

A poor correlation between ionized and total Mg was noted in critically ill patients in studies by both Barrera et al. ($r = 0.57$) *(58)*, and by Johansson and Whiss ($r = 0.59$) *(74)*. However, after magnesium supplementation, both ionized and total Mg increased: ionized by 0.05 mmol/L and total by 0.11 mmol/L *(74)*.

Fiser et al. observed that ionized magnesium was decreased in many critically ill pediatric patients who had normal total magnesium concentrations *(22)*. In this study, the use of albumin-corrected total magnesium was not reliable in estimating the ionized magnesium.

In patients undergoing cardiopulmonary bypass surgery, plasma ionized (but not total) magnesium was decreased significantly by 24 h after bypass *(32)*.

In the report by Yeh et al. *(75)* the ionized magnesium result was assumed to always be "correct." For example, if any magnesium supplementation was given based on a low total magnesium (34 patients), it was called "inappropriate administration" in 28 of the patients who had normal ionized Mg results.

The 28 with low total Mg but normal ionized Mg were considered erroneously low, and resulted in 27 additional and presumably unnecessary repeat Mg measurements and administration of 60 g of $MgSO_4$.

Variable correlations were reported by Hutton et al. *(76)*. In the general ICU population, they reported a strong correlation (r = 0.95) between ionized and total Mg, and 144 of the 185 samples fell in the same category (hypo-, normo-, or hypermagnesemia) for both ionized and total Mg based on their corresponding reference intervals of 0.49−0.71 mmol/L and 0.70−1.0 mmol/L, respectively. Also, the regression slope of 0.71 corresponded with 70% of magnesium being in the ionized form. However, in citrate-anticoagulated patients treated with continuous venovenous hemofiltration (CVVH), half of these patients showed normal total Mg and decreased ionized Mg. Based on this, they concluded that ionized Mg was better than total Mg as a marker for hypomagnesaemia in these CVVH patients.

The results from the study by Wilkes et al. could be interpreted to favor either ionized or total Mg measurements. They emphasized that correcting the ionized Mg by magnesium supplementation during cardiopulmonary bypass (CPB) reduced postoperative risks. However, their figures show that both total and ionized Mg declined almost identically during CPB and increased almost identically when magnesium supplementation was given to correct the hypo-magnesemia *(15)*.

To conclude the discussion of whether total or ionized magnesium measurement is more clinically appropriate, Longstreet and Vink provided an interesting theory *(73)*. The chronic or acute nature of the patients' condition may be relevant. In most chronic conditions, a generalized magnesium deficiency would maintain a consistent decline in both the total and ionized magnesium pools. However, in acute conditions that often require critical care, the rapid onset could rapidly alter the binding of Mg ions in serum and possibly cause greater relative differences between total and ionized magnesium concentrations. Thus, the chronic or acute nature of the disease process may be most relevant in deciding to measure total or ionized Mg to assess magnesium status.

Reference intervals for magnesium

Reference intervals (ranges) for total and ionized magnesium are shown in Table 6.4, and are based on literature sources. For ionized Mg, the lower limit of the reference intervals varied from 0.43 to 0.49 mmol/L and the upper limit varied from 0.59 to 0.71 mmol/L *(15,58,74−78)*. For total Mg, the lower limits varied from 0.64 to 0.76 mmol/L and the upper limits varied from 0.90 to 1.10 mmol/L *(1,10,15,49,58,75,76)*. Concentrations of magnesium in serum are slightly higher in older children and adults.

The reference intervals for both ionized and total Mg are important because many studies categorize patients' Mg status based on the lower and upper

TABLE 6.4 Reference ranges for magnesium.

Total serum magnesium (serum)	
Newborns	0.50–0.90 mmol/L (1.22–2.19 mg/dL)
Adults	0.70–1.05 mmol/L (1.58–2.55 mg/dL)
Ionized magnesium (blood)	0.47–0.62 mmol/L (0.97–1.52 mg/dL)
Erythrocytes	1.65–2.65 mmol/L (4.01–6.44 mg/dL)
Cerebrospinal fluid	1.0–1.40 mmol/L (2.43–3.40 mg/dL)
Urine	1–5 mmol/d

reference limits of these tests. For example, increasing the lower limit for serum magnesium will increase the incidence of hypomagnesemia in these patients. One report suggests that the standard lower limit of 0.70 mmol/L will miss hypomagnesemia in some patients, and suggested that increasing the lower reference limit to the interval of 0.70–0.80 mmol/L would miss fewer cases of hypomagnesemia *(79)*, although this would also increase the rate of false-positive diagnosis of hypomagnesemia.

Self-assessment and mastery

Self-assessment questions

1. Which drug is least likely to be associated with hypomagnesemia?
   a. aminoglycosides
   b. laxatives
   c. amphotericin
   d. cyclosporin
   e. proton pump inhibitors
2. Which clinical condition is most commonly associated with hypermagnesemia?
   a. increased dietary intake
   b. hypoaldosteronism
   c. metabolic acidosis
   d. renal failure
   e. hypokalemia
3. Hypomagnesemia is NOT associated with which of the following conditions?
   a. during cardiopulmonary bypass surgery
   b. hypoparathyroidism
   c. critical illness

    d. alcoholism

    e. insulin administration

4. What is the approximate distribution of Mg in blood? *AB*, anion-bound Mg; *PB*, protein-bound Mg.

    a. 25% PB, 10% AB, 65% ionized

    b. 10% PB, 90% ionized

    c. 50% PB, 50% ionized

    d. 45% PB, 25% AB, 30% ionized

5. Where is most Mg ion reabsorbed in the kidney?

    a. collecting duct

    b. distal tubules

    c. proximal tubules

    d. ascending limb of the loop

    e. glomeruli

6. Which factor is associated with increased Mg ion reabsorption in the kidney?

    a. hyperkalemia

    b. loop diuretics

    c. TRPM6

    d. decreased GFR

    e. hypervolemia

7. Claudins are which if the following:

    a. genes that regulate TRPM6 channels

    b. genes located in the nephron DCT

    c. proteins that bind Mg ions in blood

    d. membrane proteins that control ion flow in tight junctions between cells

    e. proteins that exchange Na and Mg ions

Answer Key:

1. b

2. d

3. e

4. a

5. d

6. c

7. d

Self-assessment cases

Assess the clinical status of the following patients and their laboratory results for possible disorders of magnesium metabolism.

Case 1. A 24-year-old woman was admitted for treatment of systemic lupus erythematosus. She complained of a weight gain of ~2.3 kg during the past week and a bloated feeling. She was started on the diuretic furosemide, 20 mg every 4 h. Twenty-four hours later the patient complained of numbness and tingling in her face, hands, and feet, which rapidly progressed to acute, painful carpal spasm. Lab results at this time were as follows:

Total calcium	2.05 mmol/L (8.2 mg/dL)
Ionized calcium	1.10 mmol/L (4.4 mg/dL)
Total magnesium	0.50 mmol/L (1.2 mg/dL)

The patient was treated with 1 g of $MgSO_4$ and 0.1 g of calcium gluconate given i.v. over 10 min. Within 60 min, the patient was free of all numbness, tingling, and spasm. Lab results 2 h later were as follows:

Total calcium	2.50 mmol/L (10.0 mg/dL)
Ionized calcium	1.30 mmol/L (5.2 mg/dL)
Total magnesium	2.50 mmol/L (6.1 mg/dL)

Adapted from a case history provided by Ronald J. Elin, Chief, Clinical Pathology Department, Clinical Center, National Institutes of Health, Bethesda, MD.

Case 2. A 71-year-old female was referred to the endocrinology clinic after years of unresolved hypomagnesemia despite several hospital admissions to receive IV magnesium. She has type 2 diabetes and among her regular medications are esomeprazole, metformin, and supplements of Ca, Mg, and vitamin D. Her abnormal laboratory results (with reference intervals) were as follows:

Magnesium: 0.21 mmol/L(0.70 − 1.0)

Calcium: 1.71 mmol/L(2.20 − 2.65)

Albumin: 3.5 g/dL(3.5 − 5.2)

Urine magnesium: not detected(0.2 − 5.0 mmol/24 h)

What might be the most appropriate course of treatment?

a. take a laxative to increase serum magnesium
b. increase the levels and frequency of IV calcium and magnesium
c. stop taking the PPI esomeprazole
d. begin a K-sparing diuretic
e. decrease alcohol consumption

Case 3. A 20-day-old female child presented with seizures and hypo-magnesemia. She had no history of fever or diarrhea. Her abnormal laboratory results (with reference intervals) were as follows:

Serum magnesium: 0.35 mmol/L$(0.70 - 1.0)$

Serum calcium: 2.10 mmol/L$(2.20 - 2.55)$

PTH: 12 ng/L$(14 - 70)$

Genetic investigations eventually revealed a mutation of the gene coding for the Mg ion transporter TRPM6.

What course of treatment would be most appropriate?

a. administering nucleic acids to her cells to correct the TRPM6 mutation
b. using stem cell gene therapy to make cells that produce functional TRPM6
c. taking long-term high dose oral $MgSO_4$
d. IV infusion of bovine PTH

Case 4. A 57-year-old male was admitted because of spasms in his hands and legs, numbness in his lips, hands, and feet, and general malaise. He also reported diarrhea that had lasted a "long time." Laboratory results diagnosed hypomagnesemia, hypocalcemia, and hypokalemia, as follows:

Serum magnesium: 0.33 mmol/L$(0.70 - 1.0)$

Serum calcium: 1.97 mmol/L$(2.20 - 2.55)$

Serum potassium: 3.4 mmol/L$(3.6 - 5.0)$

His history included coronary artery disease, hypertension, gastrointestinal reflux (GERD), and normal alcohol consumption. He was diagnosed with type-2 diabetes about 12 years ago and has been on metformin for about the last 9 years.

What would be the most likely cause of these laboratory results?

a. primary hypoparathyroidism
b. excess alcohol intake causing renal loss of Mg
c. insulin deficiency causing hypomagnesemia and hypocalcemia.
d. poor intestinal absorption of Mg due to diarrhea induced by long-term metformin treatment.
e. hypomagnesemia induced hypoparathyroidism

Case 5. A 56-year-old female was evaluated in the ED by cardiology for recurrent palpitations from supraventricular arrhythmias. She was given a calcium-blocking agent (diltiazem) and an angiotensin-converting enzyme inhibitor (enalapril). Her history also included hypertension, type-2 diabetes, and a family history of cardiac disease. For her diabetes, she has been on metformin. While most of her laboratory results were normal, her serum Mg was decreased at 1.4 mg/dL (RI = 1.7–2.7 mg/dL), her urinary excretion of Mg was slightly high, and she had a decreased vitamin D concentration. For these, she was treated with IV Mg solution and given oral Mg and vitamin D supplements. Following this, her serum Mg concentration normalized and she has had only rare episodes of palpitations.

What do you think is a likely cause of her hypomagnesemia? Select all appropriate options.

a. primary hypoparathyroidism
b. diabetic nephropathy causing renal loss of Mg
c. drug-induced by metformin
d. vitamin D deficiency
e. alcoholism

Self-assessment and mastery discussion

Assessment of Case 1. The patient experienced an acute rapid change in serum magnesium concentration, which caused more severe clinical symptoms than seen with a more gradual change. The hypomagnesemia was accompanied by a mild hypocalcemia. The cause of the rapid decrease in magnesium was the repeated doses of furosemide. Furosemide is a diuretic that acts primarily on the ascending loop of Henle to increase urinary loss of magnesium. Administration i.v. of Mg and Ca relieved the symptoms of numbness and tingling. The elevated magnesium after treatment should quickly decline with renal excretion.

Assessment of Case 2. This was a case of hypomagnesemia induced by long-term use of the PPI esomeprazole. The patient stopped this drug and was prescribed ranitidine. After this, her magnesium, calcium, and phosphate recovered within the reference intervals (From *Clin. Chem.* **2015,** *61*(5), 699–703).

Assessment of Case 3. (c) The lack of functional TRPM6 ion transporters was preventing normal intestinal absorption and renal reabsorption of Mg. By giving continual high-dose oral magnesium supplements, at 4 years old she remains well with no further seizures despite a low serum magnesium in the range of 0.4–0.6 mmol/L (From *F1000Research* **2019,** *8,* 666–684).

Assessment of Case 4. (d) The diagnosis was that the patient's severe hypomagnesemia was most likely due to the long history of diarrhea caused by the long-term use of metformin. While this is not a common cause of hypomagnesemia, that was the most likely cause because soon after he discontinued the metformin and began daily supplements of magnesium and potassium, his magnesium, calcium, and potassium were all within reference intervals (from *Cases J.* **2009,** *2,* 156–159).

Assessment of Case 5. (b, c, and d) The hypomagnesemia along with the slightly elevated urine Mg indicated that a diabetic nephropathy was the most likely cause, which was made worse by the vitamin D deficiency. She had no history of alcoholism and no evidence of hypoparathyroidism. This case illustrates the importance of measuring serum Mg whenever symptoms suggest hypomagnesemia may be a cause (adapted from Smith, N. E.; Zack, P. M. Hypomagnesemia: A Case Study. Advance for Medical Laboratory Professionals. *Immunol./Serol.* http://www.iacld.ir/DL/elm/hypomagnese miaacasestudy).

References

1. Hansen, B.-A.; Bruserud, Ø. Hypomagnesemia in Critically Ill Patients. *J. Intensive Care.* **2018,** *6,* 21.

2. Blaine, J.; Chonchol, M.; Levi, M. Renal Control of Calcium, Phosphate, and Magnesium Homeostasis. *Clin. J. Am. Soc. Nephrol.* **2015,** *10,* 1257−1272.

3. Toffaletti, J.; Cooper, D.; Lobaugh, B. The Response of Parathyroid Hormone to Specific Changes in Either Ionized Calcium, Ionized Magnesium, or Protein-Bound Calcium in Humans. *Metabolism* **1991,** *40,* 814−818.

4. Wang, S.; McDonnell, B.; Sedor, F. S.; Toffaletti, J. G. pH Effects on Measurement of Ionized Calcium and Ionized Magnesium in Blood. *Clin. Chim. Acta* **2002,** *126,* 947−950.

5. Quamme, G. A.; de Rouffignac, C. Epithelial Magnesium Transport and Regulation by the Kidney. *Front. Biosci.* **2000,** *5,* D694−D711.

6. Lisanne, M. M.; Gommers, L. M. M.; Hoenderop, J. G. J.; Bindels, R. J. M.; de Baaij, J. H. F. Hypomagnesemia in Type 2 Diabetes: A Vicious Circle? *Diabetes* **2016,** *65,* 3−13.

7. Workinger, J. L.; Doyle, R. P.; Bortz, J. Challenges in the Diagnosis of Magnesium Status. *Nutrients* **2018,** *10,* 1202. https://doi.org/10.3390/nu10091202.

8. Noronha, J. L.; Matuschak, G. M. Magnesium in Critical Illness: Metabolism, Assessment, and Treatment. *Intensive Care Med.* **2002,** *28,* 667−679.

9. Topf, J. M.; Murray, P. T. Hypomagnesemia and Hypermagnesemia. *Rev. Endocr. Metab. Disord.* **2003,** *4,* 195−206.

10. Rodriguez-Moran, M.; Guerrero-Romero, F. Oral Magnesium Supplementation Improves Insulin Sensitivity and Metabolic Control in Type 2 Diabetic Subjects. *Diabetes Care* **2003,** *26,* 1147−1152.

11. Fulop, T.; Batuman, V. *Hypomagnesemia;* eMedicine, October 2020. https://emedicine. medscape.com/article/2038394.

12. Chaigne-Delalande, B.; Lenardo, M. J. Divalent Cation Signaling in Immune Cells. *Trends Immunol.* **2014,** *35,* 332−344.

13. Gomez, M. N. Magnesium and Cardiovascular Disease. *Anesthesiology* **1998,** *89,* 222−240.

14. Pearson, P. J.; Evora, P. R. B.; Seccombe, J. F.; Schaff, H. V. Hypomagnesemia Inhibits Nitric Oxide Release from Coronary Endothelium: Protective Role of Magnesium Infusion after Cardiac Operations. *Ann. Thorac. Surg.* **1998,** *65,* 967−972.

15. Wilkes, N. J.; Mallett, S. V.; Peachey, T.; et al. Correction of Ionized Plasma Magnesium during Cardiopulmonary Bypass Reduces Risk of Postoperative Cardiac Arrhythmia. *Anesth. Analg.* **2002,** *95,* 828−834.

16. White, J. R., Jr.; Campbell, R. K. Magnesium and Diabetes: A Review. *Ann. Pharmacother.* **1993,** *27,* 775−780.

17. Salem, M.; Munoz, R.; Chernow, B. Hypomagnesemia in Critical Illness. *Crit. Care Clin.* **1991,** *7,* 225−252.

18. Rubeiz, G. J.; Thill-Baharozian, M.; Hardie, D.; Carlson, R. W. Association of Hypomagnesemia and Mortality in Acutely Ill Medical Patients. *Crit. Care Med.* **1993,** *21,* 203−209.

19. Lum, G. Hypomagnesemia in Acute and Chronic Care Patient Populations. *Am. J. Clin. Pathol.* **1992**, *97*, 827−830.

20. Escuela, M. P.; Guerra, M.; Anon, J. M.; et al. Total and Ionized Magnesium in Critically Ill Patients. *Intensive Care Med.* **2005**, *31*, 151−156.

21. Broner, C. W.; Stidham, G. L.; Westenkirchner, D. F.; et al. Hypermagnesemia and Hypocalcemia as Predictors of High Mortality in Critically Ill Pediatric Patients. *Crit. Care Med.* **1990**, *18*, 921−928.

22. Fiser, R. T.; Torres, A.; Butch, A. W.; Valentine, J. L. Ionized Magnesium Concentrations in Critically Ill Children. *Crit. Care Med.* **1998**, *26*, 2048−2052.

23. Lanzinger, M. J.; Moretti, E. W.; Toffaletti, J. G.; et al. The Relationship between Ionized and Total Serum Magnesium Concentrations during Abdominal Surgery. *J. Clin. Anesth.* **2003**, *15*, 245−249.

24. Koch, S. M.; Warters, R. D.; Mehlhorn, U. The Simultaneous Measurement of Ionized and Total Calcium and Ionized and Total Magnesium in Intensive Care Unit Patients. *J. Crit. Care* **2002**, *17*, 203−205.

25. Huijgen, H. J.; Soesan, M.; Sanders, R.; et al. Magnesium Levels in Critically Ill Patients. What Should We Measure? *Am. J. Clin. Pathol.* **2000**, *114*, 688−695.

26. Malon, A.; Brockman, C.; Fijalkowska-Morawska, J.; Rob, P.; Maj-Zurawska, M. Ionized Magnesium in Erythrocytes − The Best Magnesium Parameter to Observe Hypo- or Hypermagnesemia. *Clin. Chem. Acta* **2004**, *349*, 67−73.

27. Stalnikowicz, R. The Significance of Routine Serum Magnesium Determination in the ED. *Am. J. Emerg. Med.* **2003**, *21*, 444−447.

28. Maier, J. A. M.; Malpuech-Brugere, C.; Zimowska, W.; et al. Low Magnesium Promotes Endothelial Cell Dysfunction: Implications for Atherosclerosis, Inflammation and Thrombosis. *Biochim. Biophys. Acta* **2004**, *1689*, 13−21.

29. Rude, R.; Manoogian, C.; Ehrlich, L.; et al. Mechanism of BP Regulation by Mg in Man. *Magnesium* **1989**, *8*, 266−273.

30. Gu, W. J.; Wu, Z. J.; Wang, P. F.; Aung, L. H.; Yin, R. X. Intravenous Magnesium Prevents Atrial Fibrillation after Coronary Artery Bypass Grafting: A Meta-analysis of 7 Double-Blind, Placebo-Controlled, Randomized Clinical Trials. *Trials* **2012**, *13*, 41.

31. Cavell, G. F.; Bryant, C.; Jheeta, S. Iatrogenic Magnesium Toxicity Following Intravenous Infusion of Magnesium Sulfate: Risks and Strategies for Prevention. *BMJ Case Rep.* **2015**, *2015*. https://doi.org/10.1136/bcr-2015-209499.

32. Brookes, C. I. O.; Fry, C. H. Ionised Magnesium and Calcium in Plasma from Healthy Volunteers and Patients Undergoing Cardiopulmonary Bypass. *Br. Heart J.* **1993**, *69*, 404−408.

33. Rasmussen, H. S.; Suenson, M.; McNair, P.; Norregard, P.; Balslev, S. Magnesium Infusion Reduces the Incidence of Arrhythmias in Acute Myocardial Infarction. A Double-Blind Placebo-Controlled Study. *Clin. Cardiol.* **1987**, *10*, 351−356.

34. Woods, K. L.; Fletcher, S. Long-term Outcome after Intravenous Magnesium Sulphate in Suspected AMI: The Second Leicester Intravenous Magnesium Intervention Trial (LIMIT-2). *Lancet* **1994**, *343*, 816−819.

35. Schechter, M.; Kaplinsky, E.; Rabinowitz, B. The Rationale of Magnesium Supplementation in Acute Myocardial Infarction. A Review of the Literature. *Arch. Intern. Med.* **1992**, *152*, 2189−2196.

36. Early Administration of Intravenous Magnesium to High-Risk Patients with Acute Myocardial Infarction in the MAGIC Trial: A Randomized Controlled Trial. *Lancet* **2002**, *360*, 1189−1196.

37. Shechter, M. Magnesium and Cardiovascular System. *Magnes. Res.* **2010,** *23,* 60–72.
38. Dorman, B. H.; Sade, R. M.; Burnette, J. S.; et al. Magnesium Supplementation in the Prevention of Arrhythmias in Pediatric Patients Undergoing Surgery for Congenital Heart Defects. *Am. Heart J.* **2000,** *139,* 522–528.
39. Atsmon, J.; Dolev, E. Drug-induced Hypomagnesemia: Scope and Management. *Drug Saf.* **2005,** *28,* 763–788.
40. Garnier, A. S.; Duveau, A.; Planchais, M.; Subra, J. F.; Sayegh, J.; Augusto, J. F. Serum Magnesium after Kidney Transplantation: A Systematic Review. *Nutrients* **June 6, 2018,** *10* (6), 729.
41. Kao, W. H. L.; Folsom, A. R.; Nieto, F. J.; et al. Serum and Dietary Magnesium and the Risk for Type 2 Diabetes Mellitus. The Atherosclerosis Risk in Communities Study. *Arch. Intern. Med.* **1999,** *159,* 2151–2159.
42. Huerta, M. G.; Roemmich, J. N.; Kington, M. L.; et al. Magnesium Deficiency is Associated with Insulin Resistance in Obese Children. *Diabetes Care* **2005,** *28,* 1175–1181.
43. Dewitte, K.; Dhondt, A.; Giri, M.; et al. Differences in Serum Ionized and Total Magnesium Values during Chronic Renal Failure between Nondiabetic and Diabetic Patients. *Diabetes Care* **2004,** *27,* 2503–2505.
44. Pham, P. C.; Pham, P. M.; Pham, P. A.; et al. Lower Serum Magnesium Levels are Associated with More Rapid Decline of Renal Function in Patients with Diabetes Mellitus Type 2. *Clin. Nephrol.* **2005,** *63,* 429–436.
45. American Diabetes Association. Magnesium Supplementation in the Treatment of Diabetes. *Diabetes Care* **1992,** *15,* 1065–1067.
46. Evert, A. B.; Boucher, J. L.; Cypress, M.; Dunbar, S. A.; Franz, M. J.; Mayer-Davis, E. J.; et al. Nutrition Therapy Recommendations for the Management of Adults with Diabetes. *Diabetes Care* **2013,** *36,* 3821–3842.
47. Flink, E. B. Magnesium Deficiency in Alcoholism. *Alcohol Clin. Exp. Res.* **1986,** *10,* 590–594.
48. Hristova, E. N.; Rehak, N. N.; Cecco, S.; et al. Serum Ionized Magnesium in Chronic Alcoholism: Is It Really Decreased? *Clin. Chem.* **1997,** *43,* 394–399.
49. Yu, A. S. L.; Gupta, A. *Hypermagnesemia: Causes, Symptoms, and Treatment.* UpToDate, August 2020. www.uptodate.com/hypermagnesemia-causes-symptome-and-treatment.
50. Zafar, M. S. H.; Wani, J. I.; Karim, R.; Mir, M. M.; Koul, P. A. Significance of Serum Magnesium Levels in Critically Ill Patients. *Int. J. Appl. Basic Med. Res.* **2014,** *4* (1), 34–37.
51. Wei, M.; Esbaei, K.; Bargman, J.; Oreopoulos, D. G. Relationship between Serum Magnesium, Parathyroid Hormone, and Vascular Calcification in Patients on Dialysis: A Literature Review. *Perit. Dial. Int.* **2006,** *26,* 366–373.
52. Huang, C. L.; Kuo, E. Mechanism of Hypokalemia in Magnesium Deficiency. *J. Am. Soc. Nephrol.* **2007,** *18,* 2649–2652.
53. Rude, R. K.; Oldham, S. B.; Singer, F. R. Functional Hypoparathyroidism and Parathyroid Hormone End-Organ Resistance in Human Magnesium Deficiency. *Clin. Endocrinol.* **1976,** *5,* 209–224.
54. Hypomagnesemia. www.emedicine.com/emerg/TOPIC274.HTM. (accessed November 2008).
55. Papazacharion, I. M.; Martinez-Isla, A.; Efthimiou, E.; et al. Magnesium Deficiency in Patients with Chronic Pancreatitis Identified by an Intravenous Loading Test. *Clin. Chim. Acta* **2000,** *302,* 145–154.
56. Agus, Z. S. Hypomagnesemia. *J. Am. Soc. Nephrol.* **1999,** *10,* 1616–1622.

57. Oster, J. R.; Epstein, M. Management of Magnesium Depletion. *Am. J. Nephrol.* **1988,** *8,* 349−354.

58. Barrera, R.; Fleisher, M.; Groeger, J. Ionized Magnesium Supplementation in Critically Ill Patients: Comparing Ionized and Total Magnesium. *J. Crit. Care* **2000,** *15,* 36−40.

59. Yamamoto, M.; Yamaguchi, T. Causes and Treatment of Hypomagnesemia. *Clin. Calcium* **2007,** *17,* 1241−1248.

60. Velissaris, D.; Karamouzos, V.; Pierrakos, C.; Aretha, D.; Karanikolas, M. Hypomagnesemia in Critically Ill Sepsis Patients. *J. Clin. Med. Res.* **2015,** *7,* 911−918.

61. Felsenfeld, A. J.; Levine, B. S.; Rodriguez, M. Pathophysiology of Calcium, Phosphorus, and Magnesium Dysregulation in Chronic Kidney Disease. *Semin. Dial.* **2015,** *28,* 564−577.

62. Weiss-Guillet, E.-M.; Takala, J.; Jakob, S. M. Diagnosis and Management of Electrolyte Emergencies. *Best Pract. Res. Clin. Endocrinol. Metab.* **2003,** *17,* 623−651.

63. Monif, G. R.; Savory, J. Iatrogenic Maternal Hypocalcemia Following Magnesium Sulfate Therapy. *J. Am. Med. Assoc.* **1972,** *219,* 1469.

64. Abassi-Ghanavati, M.; Alexander, J. M.; McIntire, D. D.; et al. Neonatal Effects of Magnesium Sulfate Given to the Mother. *Am. J. Perinatol.* **2012,** *29,* 795.

65. Horino, T.; Ichii, O.; Terada, Y. A Rare Presentation of Hypermagnesemia Associated with Acute Kidney Injury Due to Hypercalcemia. *Intern. Med.* **2019,** *58,* 1123.

66. Castelbaum, A. R.; Donofrio, P. D.; Walker, F. O.; Troost, B. T. Laxative Abuse Causing Hypermagnesemia, Quadriparesis, and Neuromuscular Junction Defect. *Neurology* **1989,** *39,* 746.

67. Schelling, J. R. Fatal Hypermagnesemia. *Clin. Nephrol.* **2000,** *53,* 61−65.

68. Karosanidze, T. Magnesium − So Underappreciated. *Pract. Gastroenterol.* **January 2014,** *38.*

69. Rouilly, M. V.; Badertscher, M.; Pretsch, E.; Suter, G.; Simon, W. Neutral-Carrier-Based Magnesium-Selective Electrode. *Anal. Chem.* **1988,** *60,* 2013−2016.

70. Cecco, S. A.; Hristova, E. N.; Rehak, N. N.; Elin, R. J. Clinically Important Intermethod Differences for Physiologically Abnormal Ionized Magnesium Results. *Am. J. Clin. Pathol.* **1997,** *108,* 564−569.

71. Saha, H.; Harmoinen, A.; Karvonen, A. L.; et al. Serum Ionized Versus Total Magnesium in Patients with Intestinal or Liver Disease. *Clin. Chem. Lab. Med.* **1998,** *36,* 715−718.

72. Saha, H.; Harmoinen, A.; Pietila, K.; et al. Measurement of Serum Ionized Versus Total Levels of Magnesium and Calcium in Hemodialysis Patients. *Clin. Nephrol.* **1996,** *46,* 326−331.

73. Longstreet, D.; Vink, R. Correlation between Total and Ionic Magnesium Concentration in Human Serum Samples is Independent of Ethnicity or Diabetic State. *Magnes. Res.* **2009,** *22* (1), 32−36.

74. Johansson, M.; Whiss, P. A. Weak Relationship between Ionized and Total Magnesium in Serum of Patients Requiring Magnesium Status. *Biol. Trace Elem. Res.* **2007,** *115,* 13−21.

75. Yeh, D. D.; Chokengarmwong, N.; Chang, Y.; Yu, L.; Arsenault, C.; Rudolf, J.; et al. Total and Ionized Magnesium Testing in the Surgical Intensive Care Unit - Opportunities for Improved Laboratory and Pharmacy Utilization. *J. Crit. Care* **2017,** *42,* 147−151.

76. Hutten, T. J. A.; Sikma, M. A.; Stokwielder, R. H.; Wesseling, M.; Imo, E.; Hoefer, I. E.; et al. Ionized and Not Total Magnesium as a Discriminating Biomarker for Hypomagnesaemia in Continuous Venovenous Haemofiltration Patients. *Nephrol. Dial. Transplant.* **2021,** 1−3. https://doi.org/10.1093/ndt/gfaa330.

77. Munoz, R.; Laussen, P. C.; Palacia, G.; Zienko, L.; Piercey, G.; Wessel, D. L. Whole Blood Ionized Magnesium: Age-Related Differences in Normal Values and Clinical Implications of

Ionized Hypomagnesemia in Patients Undergoing Surgery for Congenital Cardiac Disease. *J. Thorac. Cardiovasc. Surg.* **2000,** *119,* 891−898.

78. Hristova, E. N.; Cecco, S.; Niemela, J. E.; Rehak, N. N.; Elin, R. J. Analyzer-Dependent Differences in Results for Ionized Calcium, Ionized Magnesium, Sodium, and pH. *Clin. Chem.* **1995,** *41,* 1649−1653.

79. Leibscher, D.-H.; Liebscher, D.-E. About the Misdiagnosis of Magnesium Deficiency. *J. Am. Coll. Nutr.* **2004,** *23,* 730S.

Chapter 7

Phosphate physiology and clinical evaluation

Introduction

Symptoms of phosphate depletion have been described through the years, from the times of the ancient Romans to observations by veterinarians on livestock. Phosphate compounds are in all cells and participate in numerous biochemical processes. Phosphate is a component of DNA and RNA, phospholipids, high-energy compounds such as ATP and creatine phosphate, and many coenzymes.

When both phosphate depletion and hypophosphatemia are present, serious biochemical abnormalities often result: impaired myocardial function, respiratory muscle paralysis, central nervous system disorders, and skeletal muscle degradation [1,2]. These may result from hypophosphatemia inhibiting mitochondrial respiration and ATP synthesis, processes that are vital to cellular survival. Furthermore, insulin resistance may be a consequence of hypophosphatemia [3].

More recently, several Na-coupled PO_4 transport proteins, called cotransporters, have been described that are key factors in PO_4 regulation. These cotransporters have been designated as Type 1, Type 2 (Npt2a, Npt2b, and Npt2c), and Type 3 cotransporters (PiT1 and PiT2) [1].

Distribution in cells and blood

About 80% of phosphate is in bone, mostly in the form of hydroxyapatite $[Ca_{10}(PO_4)_6(OH)_2]$. Phosphate in blood is either absorbed from dietary sources or resorbed from the bone. Most phosphate in the body outside of the bones is within cells, where the intracellular PO_4 concentration is about 100 mmol/L, most of which is either bound or complexed to proteins or lipids [1].

The total concentration of phosphate in the blood is ~ 3.9 mmol/L (~ 12 mg/dL), with most as organic phosphates. The concentration of inorganic PO_4 is about 1.0−1.3 mmol/L (3−4 mg/dL), all of which is completely ionized, circulating primarily as hydrogen phosphate HPO_4^{-2} or dihydrogen phosphate $H_2PO_4^-$ in a ratio of 4:1 at plasma pH of 7.40 [1,4].

Blood Gases and Critical Care Testing. https://doi.org/10.1016/B978-0-323-89971-0.00007-0

Physiology and regulation

Cellular transport

Although cellular shifts of phosphate can affect phosphate concentrations in the blood and ECF, absorption from the intestine and excretion by the kidney are the dominant homeostatic mechanisms. Because absorption by the intestine fluctuates widely, the kidney is responsible for the precise regulation of phosphate concentrations in the blood (5).

Cotransporters are carrier proteins that are essential in PO_4 regulation. Cotransporters simultaneously transport two different molecules or ions from one side of the membrane to the other, amazingly at a rate of several thousand molecules per second. By using a favorable electrical or concentration gradient to transport one ion across a membrane, another ion may be transported against its concentration gradient. Because the intracellular concentration of phosphate is greater than the extracellular concentration, phosphate entry into cells requires a facilitated transport process. An example is the transport of a Na ion from high to low concentration provides the energy to transport a PO_4 ion intracellularly against a concentration gradient.

Several Na-coupled transport proteins enable intracellular uptake of PO_4 by taking advantage of the steep extracellular-to-intracellular Na gradient (1). The actions of these cotransporters in the nephron are shown in Fig. 7.1.

Type 1 Na/PO_4 cotransporters are expressed predominantly in kidney proximal tubule cells, although their role in phosphate homeostasis is not clear.

Type 2a cotransporters (Npt2a) in the kidney proximal convoluted tubules (PCTs) are responsible for absorbing most of the filtered phosphate. To maintain PO_4 homeostasis, their expression is increased when PO_4 is needed and their expression is decreased by high dietary phosphate intake, parathyroid hormone (PTH), fibroblast growth factor 23 (FGF23), and dopamine.

Type 2b cotransporters (Npt2b) are similar to type 2a transporters, are expressed in the small intestine, and are also upregulated under conditions of dietary phosphate deprivation.

Type 2c Na/PO_4 cotransporters (Npt2c), together with Type 2a transporters, are essential for normal phosphate homeostasis, and are also regulated by diet and PTH. Loss of type 2c function can result in hypophosphatemic rickets (1).

Type 3 Na/PO_4 cotransporters (PiT1 and PiT2) are in almost all cells and presumably play a housekeeping role in ensuring adequate phosphate for all cells.

Serum levels of phosphate are affected by intestinal PO_4 absorption mediated by the Npt2b cotransporter. Npt2b is regulated by both dietary phosphate intake and $1,25(OH)_2$ vitamin D, with most intestinal phosphate reabsorbed in the duodenum and the jejunum.

Maintenance of normal serum phosphorus levels is closely regulated by PO_4 reabsorption within the nephron, where approximately 85% of phosphate is reabsorbed within the proximal tubule. The remainder of the nephron plays a minor role in PO_4 regulation (5).

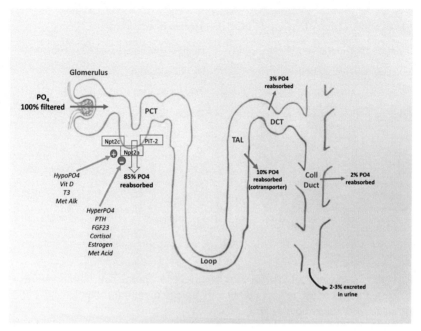

FIGURE 7.1 Processing of PO_4 ions in different segments of the nephron. This shows the 100% filtration of PO_4 ions through the glomerulus and the reabsorption in the proximal convoluted tubule (PCT), the thick ascending limb (TAL) of the loop, the distal convoluted tubule (DCT), and the collecting duct (Coll Duct). Abbreviations used: *FGF23*, fibroblast growth factor 23; *HyperPO4*, hyperphospatemia; *HypoPO4*, hypophosphatemia; *Met Alk*, metabolic alkalosis; *Npt2a*, type 2a Na/PO_4 cotransporters; *Npt2c*, type 2c Na/PO_4 cotransporters; *PiT-2*, type 3 Na/PO_4 cotransporters; *PTH*, parathyroid hormone; *T3*, tri-iodothyronine; *Vit D*, vitamin D. *This is an original drawing.*

Within the PCT, PO_4 transport from the ultrafiltrate across the proximal tubule epithelium is an energy-dependent process that utilizes three Na/PO_4 cotransporters, Npt2a, Npt2c, and PiT-2 located in the PCT cells. Electrical potentials generated from the transport of Na ions facilitate the movement of PO_4 ions from the proximal tubular filtrate into the cells. The amount of PO_4 reabsorbed from the filtrate is determined by the number of these cotransporters expressed in the PCT cells and not by alterations in their ability to transport PO_4.

Factors that affect renal absorption of PO_4

Several factors that affect PO_4 absorption in the kidney are listed in Table 7.1.

Parathyroid hormone (PTH). PTH decreases renal reabsorption of phosphate by decreasing the abundance of the Na/PO_4 cotransporters Npt2a, Npt2c, and PiT-2 in the renal proximal tubule brush border membrane. In response to

TABLE 7.1 Factors that affect PO_4 absorption in the kidney *(5)*.

Factors that increase PO_4 absorption	Factors that decrease PO_4 absorption
Dietary deficiency of PO_4	Parathyroid hormone
1,25 $(OH)_2$-Vitamin D	Fibroblast growth factor (FGF23)
Thyroid hormone (T3)	Dietary excess of PO_4
Metabolic alkalosis	Metabolic acidosis
	Glucocorticoids
	Dopamine
	Estrogens
	Hypertension

PTH, the abundance of Npt2a cotransporters is diminished within minutes, whereas the decrease in the number of apical membrane Npt2c and PiT-2 cotransporters takes hours *(5)*.

While PTH is primarily involved in the regulation of plasma calcium concentrations, PTH also acts on phosphate. As PTH concentrations increase, the bone releases calcium and phosphate-containing minerals into circulation. Because PTH increases renal tubular reabsorption of calcium and decreases phosphate reabsorption in the proximal tubules, the net effect is that PTH decreases serum phosphate concentrations. PTH also promotes formation in the kidneys of 1,25-dihydroxycholecalciferol (calcitriol), which enhances absorption of calcium and phosphate in the GI tract *(3)*. The actions of these hormones and cotransporters in the nephron are shown in Fig. 7.1.

Fibroblast growth factor-23 (FGF23). FGF23, produced in osteoblasts in response to increases in serum PO_4, has several hypophosphatemic actions. FGF23:

- Reduces the number of Na/PO_4 cotransporters in the renal proximal tubule,
- Reduces serum levels of calcitriol by decreasing the renal expression of 1α-hydroxylase, which is the rate-limiting step in calcitriol synthesis.
- Increases renal expression of 24-hydroxylase, which minimizes calcitriol production.
- Suppresses PTH synthesis in healthy kidneys *(5)*.

Calcitriol (1,25(OH)_2-vit D). In response to hypophosphatemia, 25-hydroxycholecalciferol is converted by 1α-hydroxylase to 1,25-dihydroxycholecalciferol (calcitriol), which increases PO_4 reabsorption in the proximal. Calcitriol also regulates its own homeostasis by simultaneously suppressing 1α-hydroxylase and activating 24-hydroxylase, which makes inactive vitamin D *(5)*. Calcitriol also regulates the Npt2b cotransporter in the intestine, and it increases the activity of the osteoblasts to lay down calcium into the bone matrix *(3)*.

Glucocorticoids. Glucocorticoids decrease phosphate reabsorption by decreasing proximal tubule synthesis of Npt2a in membranes of the proximal tubules *(5)*.

Estrogens. Estrogen causes phosphaturia by decreasing the number of Npt2a cotransporters in the proximal tubule. Estrogen also increases FGF23 synthesis *(5)*.

Thyroid hormone. Thyroid hormone increases phosphate absorption by increasing proximal tubule expression of Npt2a. Npt2a gene transcription of mRNA is regulated by triiodothyronine *(5)*.

Acid−base disorders. Metabolic acidosis stimulates phosphaturia by inhibiting Na/PO_4 cotransporter activity. Metabolic alkalosis increases renal phosphate absorption, possibly by increasing the abundance of Npt2a and Npt2c *(5)*.

Hypophosphatemia

Causes of hypophosphatemia

The incidence of hypophosphatemia in hospitalized patients varies greatly, depending on the population studied *(2,6)*. For example, hypophosphatemia may occur in 1% of general patients upon admission. The percentage increases with length of stay, critical illness, and alcoholism.

Hypophosphatemia is commonly caused by transcellular shifts of phosphate, GI malabsorption, or renal loss. Dietary deficiency is rarely a cause of hypophosphatemia, both because PO_4 is present in most foods and because the intestines are highly efficient in absorbing PO_4 *(1)*.

The mechanisms of hypophosphatemia are described as follows:

Transcellular shifts. Because the movement of glucose into cells is accompanied by phosphate, the administration of glucose or insulin will lead to an influx of phosphate into cells that may cause hypophosphatemia. Respiratory alkalosis also causes hypophosphatemia because CO_2 loss increases pH, which stimulates glycolysis which enhances phosphate uptake by cells *(3)*.

GI losses. As with many electrolyte disorders, diarrhea and vomiting can diminish GI absorption of phosphate. Because the divalent cations in antacids such as aluminum hydroxide, magnesium hydroxide, or aluminum carbonate bind phosphate and prevent its absorption in the gut, hypophosphatemia often results from chronic excessive use of such antacids (3). These conditions are made worse if patients have vitamin D deficiency.

Renal losses. Renal loss of PO_4 is most commonly related to the use of diuretics. Other causes include primary hyperparathyroidism, hypomagnesemia, and defects in vitamin D metabolism *(1)*.

Other causes. Hypophosphatemia can be caused by a mix of factors:

• In diabetic ketoacidosis, the combined effects of acidosis, glycosuria, ketonuria, and insulin therapy deplete PO_4, both by renal loss and cellular uptake. Serum concentrations may decrease rapidly.

• Alcoholism leads to hypophosphatemia by increasing renal loss and decreasing GI absorption of PO_4.

• During recovery from thermal burns, PO_4 is secreted into renal tubules, where it is lost in the urine.

• Cytokines have been associated with intracellular PO_4 shifts, which may be responsible for hypophosphatemia in trauma, burns, and sepsis *(1,7)*.

Symptoms of hypophosphatemia

Severe hypophosphatemia usually results in decreased concentrations of phosphate-containing compounds, such as ATP and membrane phospholipids. These deficiencies are responsible for symptoms of hypophosphatemia, such as muscle weakness, seizures, respiratory and myocardial insufficiency, and hepatocellular damage *(1,2)*.

Hypophosphatemic patients have a marked increase in urinary calcium and magnesium excretion, caused by both increased bone loss and altered renal tubular handling of these ions. Phosphate depletion also suppresses PTH secretion, which further enhances the urinary loss of Mg and Ca ions.

Moderate hypophosphatemia can lead to weakness of pulmonary muscles *(1,2)* and can prolong the weaning of patients from a ventilator. These effects are usually reversed by phosphate repletion.

More severe hypophosphatemia is associated with life-threatening seizures, coma, and dysfunction of respiratory and myocardial muscles. Other undesirable cardiovascular effects include impaired mitochondrial oxygen consumption and decreased sensitivity to inotropic agents, such as epinephrine *(2)*.

Phosphate depletion causes a reduction in 2,3-DPG in erythrocytes, which increases the affinity of Hb for oxygen. This can reduce oxygen release to tissues by 10%−15% in severe hypophosphatemia *(2)*. If hypophosphatemia occurs suddenly, it can cause hemolysis.

Phosphate depletion causes a rapid breakdown of the bone matrix, even more so than that caused by either severe hypocalcemia or vitamin D depletion. These effects appear to be independent of vitamin D or PTH *(2)*.

Evaluation of hypophosphatemia

When hypophosphatemia is confirmed, the sequence of laboratory tests that may be helpful is shown in Fig. 7.2, and the interpretation of labortory test results in renal and non-renal causes of hypophosphatemia is shown in Table 7.2.

Laboratory Tests in Evaluating Hypophospatemia

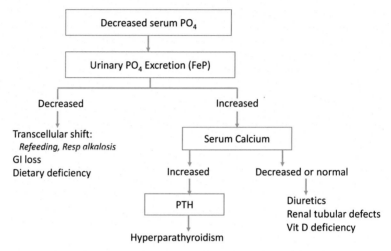

FIGURE 7.2 Use of laboratory tests in evaluation of hypophosphatemia. *ATN*, acute tubular necrosis; *vit D*, vitamin D, *FeP*, fractional excretion of PO_4. *This is an original drawing.*

If hypophosphatemia is present, a 24-h urine collection can determine if phosphate wasting is present if the fractional excretion of PO_4 is greater than about 15% *(1)*. A low urinary phosphate excretion [<3.3 mmol (or 100 mg)/d)] suggests a nonrenal cause, such as a GI loss, a dietary deficiency, or a transcellular shift of PO_4 as with insulin administration. A high urinary phosphate excretion [>3.3 mmol (or 100 mg)/d)] indicates a renal loss of phosphate caused by a renal tubular defect, use of diuretics, hyperparathyroidism, or a vitamin D defect or deficiency.

In addition to serum phosphate, serum calcium and magnesium are often helpful. High calcium levels coupled with low phosphate levels suggest primary hyperparathyroidism, while low calcium levels suggest vitamin D deficiency or malabsorption. Because of the many factors that regulate calcium independently of phosphate, serum calcium concentrations may be within reference intervals in either of these circumstances *(1)*.

If present, a low Mg concentration may also contribute to hypophosphatemia by inhibiting PTH secretion and activity of ATP-ases. Abnormal K concentrations, especially hypokalemia, may occur with hypophosphatemia in conditions such as diabetic ketoacidosis and alcoholism.

Measurement of PTH is necessary to determine if either primary or secondary hyperparathyroidism is present. A high PTH level in the presence of high calcium and low phosphate suggests primary hyperparathyroidism. A high PTH level with low or low-normal calcium and phosphate concentrations

TABLE 7.2 Laboratory tests in renal and nonrenal causes of hypophosphatemia.

	PO_4	Ca	ALP	PTH	1,25 vit D	Urine PO_4 or FeP
Renal cause						
Hyperparathyroidism	Dec	Inc	N-Inc	Inc	Inc	Inc
Rickets (HHRH)	Dec	N	Inc	Dec	Inc	Inc
Diuretics	Dec	N-Inc	N	N	N	Inc
Post liver Txp	Dec	N	N-Inc	N	N	Inc
Fanconi syndrome	Dec	N-Dec	N-Inc	N-Inc	N-Dec	Inc
Genetic cause: XLH, ADHR. TIO, postrenal Txp	Dec	N	Inc	N-Inc	N-Dec	Inc
Nonrenal cause						
GI loss or dietary deficiency	Dec	N-Dec	N-Inc	N-Inc	–	Dec
Oral PO_4 binders in CKD	Dec	N-Inc	N-Inc	N	Inc	Dec
Intracellular uptake: insulin, alkalosis, refeeding; cytokines (sepsis)	Dec	N	N	N	–	Dec

All results are for levels of serum concentrations unless noted as urine levels. *ADHR*, autosomal dominant hypophosphatemic rickets; *CKD*, chronic kidney disease; *Dec*, decreased; *FeP*, fractional excretion of PO_4; *HHRH*, hereditary hypophosphatemic rickets; *Inc*, increased; *N*, normal; *TIO*, tumor-induced osteomalacia; *Txp*, transplant; *XLH*, X-linked hypophosphatemia.
Table modified from Imel, E. A.; Econs, M. J. Approach to the Hypophosphatemic Patient. *J. Clin. Endocrinol. Metab.* **2012,** *97*(3), 696–706.

suggests secondary hyperparathyroidism. Vitamin D levels are needed to detect vitamin D deficiency, which is common, especially in geriatric or chronically ill persons.

Arterial blood gases should be ordered if respiratory alkalosis (hyperventilation) is present because it can cause hypophosphatemia.

Serum lactate, CBC with differential, and liver function tests, including serum ammonia, may be useful in selected patients to investigate other causes of hypophosphatemia, such as sepsis and hepatic encephalopathy.

Treatment of hypophosphatemia

Acute hypophosphatemia is common in hospitalized patients and can result in significant morbidity and mortality. Chronic hypophosphatemia associated with genetic or acquired renal phosphate-wasting disorders can cause abnormal growth and rickets in children and osteomalacia in adults *(8)*.

Acute hypophosphatemia may be classified by the following intervals of serum PO_4 concentrations:

- Mild: 2–2.5 mg/dL
- Moderate: 1–1.9 mg/dL
- Severe: <1 mg/dL

Acute hypophosphatemia may occur with refeeding, alcoholism, diabetic ketoacidosis, malnutrition/starvation, after surgery, or during intensive care. Replacement may be given as phosphate solutions either orally, intravenously, or in total parenteral nutrition, with the rate and amount determined by the symptoms, severity, duration of illness, and presence of comorbid conditions such as kidney failure, volume overload, hypo- or hypercalcemia, hypo- or hyperkalemia, and acid–base status. Mild/moderate acute hypophosphatemia usually can be corrected with increased dietary phosphate supplements or milk consumption *(1,4,8)*. With severe hypophosphatemia (<1.0 mg/dL or 0.3 mmol/L), solutions of phosphate may be given intravenously. Persistent hypophosphatemia often has a poor prognosis, with a mortality rate of 20% *(2)*. In some cases, correcting hypophosphatemia may also correct other electrolyte and acid–base disorders *(1,2)*.

In chronic hypophosphatemia, standard treatment includes oral phosphate supplementation and active vitamin D, and possibly other agents such as calcitonin *(8)*. Oral supplements may be helpful in genetic disorders of phosphate wasting by normalizing PO_4 concentrations and minimizing bone pain *(1)*.

Hyperphosphatemia

Causes of hyperphosphatemia

Renal excretion of PO_4 is a highly efficient process that can maintain normal plasma PO_4 concentrations, even if phosphate intake is increased to as much as 4000 mg/day (130 mmol/day). Phosphate intake above this amount usually causes only small elevations in serum phosphate concentrations as long as the intake is distributed over the course of the day. However, if an acute phosphate load is given over a few hours, transient hyperphosphatemia will occur *(9)*.

There are four general circumstances that can cause hyperphosphatemia:

- Acute or chronic kidney disease
- A primary increase in tubular phosphate reabsorption
- Extracellular shift of phosphate
- Acute phosphate load

Kidney disease with reduced GFR and reduced renal excretion of phosphate is the most common condition associated with hyperphosphatemia and is often a contributing factor with excessive intake of PO_4-containing antacids, laxatives, enemas, vitamin D, or milk *(9,10)*.

In acute or chronic kidney disease the following mechanisms that normally reduce phosphate are impaired:

- Hyperphosphatemia diminishes proximal tubular PO_4 reabsorption by suppressing Na/PO_4 cotransporters.
- PTH increases phosphate excretion by diminishing the activity of Na/PO_4 cotransporters.
- Phosphatonins such as FGF23 enhance urinary phosphate excretion by suppressing the expression of Na/PO_4 cotransporters.

Although the reduction in GFR will initially diminish phosphate filtration and excretion, phosphate balance can initially be maintained in such patients by decreasing proximal phosphate reabsorption under the influence of increased secretion of PTH and FGF23. Once the GFR falls below 20—25 mL/min, however, phosphate reabsorption is maximally suppressed, resulting in hyperphosphatemia.

Hyperphosphatemia caused by an acute extracellular shift of phosphate sometimes occurs in lactate acidosis and diabetic ketoacidosis. Metabolic acidosis promotes cellular loss of PO_4 by diminishing cellular phosphate utilization, which causes increased serum phosphate.

An acute phosphate load sufficient to overwhelm renal capacity for excretion can be derived from either endogenous or exogenous sources. Any cause of marked tissue breakdown can release intracellular PO_4 into the extracellular fluid. Examples of marked tissue breakdown include tumor lysis syndrome, rhabdomyolysis, and, rarely, marked hemolysis or transfusion of stored blood.

Tumor lysis syndrome is usually caused by cytotoxic therapy in patients with highly active tumor cell turnover, such as lymphomas or leukemias. In addition, this syndrome is also associated with increased potassium, uric acid, and urea.

In rhabdomyolysis, damaged cells release myoglobin and phosphate, along with other cell contents. The heme released in this disease can also induce acute kidney injury *(9,10)*.

Hyperphosphatemia from exogenous sources is most commonly from ingesting large amounts of phosphate-containing laxatives. This can cause severe, even fatal, hyperphosphatemia and hypocalcemia, especially in patients with underlying chronic kidney disease (CKD). GI fluid loss (i.e., from diarrhea) along with renal insufficiency can contribute to hyperphosphatemia.

Hypoparathyroidism caused by either deficient PTH secretion or renal resistance to PTH (pseudohypoparathyroidism) results in increased phosphate reabsorption and leads to hyperphosphatemia.

Bisphosphonates, primarily etidronate, can cause mild hyperphosphatemia by direct stimulation of renal phosphate reabsorption.

Fibroblast growth factor receptor inhibitors, such as erdafitinib and pemigatinib used to treat various cancers, can cause hyperphosphatemia by inhibiting FGF23 signaling. Several types of genetic mutations of the FGF23 gene can promote decreased renal phosphate excretion.

Vitamin D toxicity will increase intestinal phosphate and calcium absorption. This causes a rise in serum calcium concentration that diminishes urinary PO_4 excretion, both by inhibiting PTH secretion and by impairing kidney function, in part due to direct renal vasoconstriction.

Symptoms of hyperphosphatemia

Most patients with hyperphosphatemia are asymptomatic but may have some degree of bone or joint pain *(10)*. Acute hyperphosphatemia that develops within hours or days may be associated with the following findings *(2,9,10)*:

- Hypocalcemia, possibly with secondary hyperparathyroidism, causing muscle cramps, tetany, or perioral numbness.
- Precipitation of calcium and phosphate salts in various tissues that may cause problems in several organs: kidneys, vascular smooth muscle, eyes, joints, and skin *(1)*.
- Impaired kidney function leading to fatigue, shortness of breath, anorexia, nausea and vomiting, and sleep disorders *(10)*.
- Conductance disturbances in the heart, leading to arrhythmias and hypotension
- Impaired O_2 diffusion in the alveoli associated with shortness of breath and fatigue

Evaluation of hyperphosphatemia

With an adult reference range of 0.81–1.45 mmol/L (2.5–4.5 mg/dL), a general range for mild-to-moderate hyperphosphatemia may be 1.6–2.9 mmol/L (5–9 mg/dL), with severe hyperphosphatemia >2.9 mmol/L (9 mg/dL) *(10)*. Because hyperphosphatemia is often associated with common types of metabolic acidosis (ketoacidosis and lactate acidosis), and either acute or chronic renal failure, hyperphosphatemia is relatively common in hospitalized populations.

As shown in Fig. 7.3, further investigation of an elevated serum phosphate should include evaluation of kidney function (creatinine, cystatin C), and measurements of Na, K, Ca, and Mg.

Laboratory Tests in Evaluating Hyperphospatemia

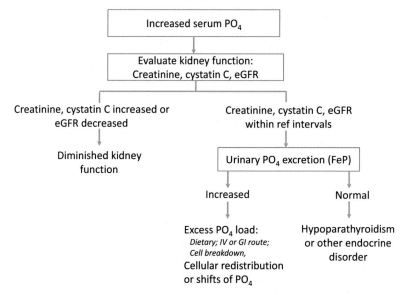

FIGURE 7.3   Use of laboratory tests in evaluation of hyperphosphatemia. *eGFR*, estimated glomerular filtration rate; *ref*, reference; *IV*, intravenous; *GI*, gastrointestinal. *This is an original drawing.*

- An elevated serum creatinine or cystatin C, or decreased eGFR calculated from them, indicates that diminished kidney function causes hyperphosphatemia.
- If tests indicate normal renal function, urinary phosphate excretion may be helpful. A normal urinary phosphate suggests increased phosphate reabsorption such as in hypoparathyroidism. An increased phosphate excretion >48 mmol (or 1500 mg)/d suggests an increased phosphate load to the kidneys such as from increased dietary intake, increased cellular breakdown (chemotherapy, rhabdomyolysis), or excess administration of phosphate by intravenous or intestinal routes *(10)*.
- Measurements of plasma PTH and vitamin D may be needed to confirm abnormalities of these hormones.

Treatment of hyperphosphatemia

Treatment should be directed at eliminating the source of phosphate, removing the excess phosphate, and correcting any associated electrolyte disorders such as hypocalcemia, hypomagnesemia, or hypokalemia *(4,8,9)*. Management depends on whether the hyperphosphatemia has occurred acutely or chronically. Acute hyperphosphatemia is serious and may require insulin and dextrose, which transfer phosphate into cells, or acetazolamide, which increases urinary excretion and rapidly normalize blood levels. If renal failure is present in serious cases of hyperphosphatemia, peritoneal dialysis or hemodialysis may be necessary *(8)*.

Chronic hyperphosphatemia commonly occurs in patients with kidney failure. It should be treated by a low-PO_4 diet to reduce intestinal absorption and by ingesting a PO_4-binding salt. In chronic hyperphosphatemia, dietary intake of phosphate should be less than 6.5 mmol (200 mg)/d *(9,10)*.

Reference intervals

Reference intervals for phosphate are shown in Table 7.3 *(10—14)*. Plasma phosphate concentrations are normally higher in children and postmenopausal women and also exhibit a diurnal rhythm, with peaks around 0400 and 1600, and with nadirs at around 0800 *(10)*.

Urinary excretion of PO_4 varies considerably according to dietary intake, age, and gender, but a reference interval of 400—1300 mg/24 h is typical *(13)*. Urine PO_4 results are often presented in proportion to the creatinine concentration. They may also be expressed as the fractional excretion of PO_4 (FeP). FeP is a measure of the fraction of the PO_4 filtered by the glomerulus, which is excreted into the urine. It is expressed as the ratio of PO_4 clearance to creatinine clearance, as shown in Table 7.3 *(14)*.

TABLE 7.3 Reference intervals for phosphate.

Plasma, serum	
Neonate (0—5 days)	1.55—2.65 mmol/L (4.8—8.2 mg/dL)
4—11 years	1.20—1.80 mmol/L (3.7—5.6 mg/dL)
12—15 years	0.95—1.75 mmol/L (2.9—5.4 mg/dL)
Adult	0.81—1.45 mmol/L (2.5—4.5 mg/dL)
Urine	
PO_4 concentration	13—28 mmol/L
PO_4 per 24 h urine volume	400—1300 mg/24 h
PO_4/Creatinine ratio	
Males	70—875 mg PO_4/g creatinine
Females	60—845 mg PO_4/g creatinine
FeP = fractional excretion of PO_4	
FeP reference interval: 10%—20%	
FeP = $([PO_4]_{UR}/[PO_4]_{PL}) \times ([Cr]_{PL}/[Cr]_{UR}) \times 100$	

Self-assessment and mastery

Self-assessment questions

1. Which clinical findings are typical of hyperparathyroidism?
   a. hyperphosphatemia
   b. hypophosphatemia
   c. hypocalcemia
   d. hypercalcemia
   e. b and d
2. Which statement is NOT correct about phosphorous homeostasis?
   a. inorganic phosphate is freely filtered by the glomerulus
   b. FGF23 promotes renal tubular reabsorption of PO4
   c. PTH increases phosphate loss in urine
   d. vitamin D increases intestinal absorption and renal reabsorption of phosphate
   e. calcitriol regulates the Npt2b cotransporter in the intestine
3. (True−False) Most phosphorous in the blood is organic phosphates.
4. Which of the following clinical conditions can cause hyperphosphatemia?
   a. both acute and chronic renal failure
   b. leukemia
   c. rhabdomyolysis
   d. decreased glomerular filtration rate
   e. all the above
5. Hypophosphatemia is NOT found in which condition?
   a. alcoholism
   b. GI maladsorption conditions
   c. respiratory alkalosis
   d. chronic renal failure
   e. excess insulin administration
6. A patient has renal failure and elevated serum phosphate and PTH levels and a chronically low serum Ca. What is the most likely diagnosis from the following choices?
   a. primary hyperparathyroidism
   b. secondary hyperparathyroidism
   c. a PTH-producing tumor
   d. tertiary hyperparathyroidism
   e. hypomagnesemia
7. What is the approximate concentration of intracellular PO_4?
   a. 4 mmol/L
   b. 10 mmol/L
   c. 25 mmol/L
   d. 100 mmol/L
   e. 200 mmol/L

8. Which of the following factors increases the number of Na/PO_4 cotransporters?
   a. glucocorticoids
   b. thyroid hormone
   c. estrogens
   d. metabolic acidosis
   e. PTH

9. True-False: Hyperphosphatemia suppresses Na/PO_4 cotransporters which diminishes tubular reabsorption of PO_4.

10. Calculate the Fractional Excretion of PO_4 (FeP) for Examples 10a and 10b using the following data:

Parameter	Example 10a	Example 10b
Serum PO_4	1.2 mmol/L	1.2 mmol/L
24 h Urine PO_4	18 mmol	36 mmol/L
Serum creatinine	1.0 mg/dL	1.0 mg/dL
24 h Urine creatinine	80 mg/dL	80 mg/dL
FeP		

Answer Key:

1. e
2. b
3. True
4. e
5. d
6. b
7. d
8. b
9. True
10a. FeP = 18.75%; 10b. FeP = 37.5%

Self-assessment cases

Assess the clinical status of the following patients and their laboratory results for possible disorders of phosphate metabolism.

   Case 1. A 52-year-old man was admitted for abdominal pain and vomiting. He had a normal weight, no history of surgery, GI disorders, or significant alcohol use. On admission, his amylase, lipase, and other serum enzymes (AST, ALT, ALP) were elevated and his phosphate and calcium were slightly decreased. His blood gases indicated a degree of respiratory alkalosis (pH 7.49; pCO_2 30 mmHg). An ultrasound showed gallstones and dilated biliary ducts. After 3 days, his serum phosphate became very low at 0.9 mg/dL.

Intravenous phosphate was given during the next few days until the hypo-phosphatemia resolved. What condition do the clinical and laboratory findings indicate and how does this relate to the phosphate concentration?

Case 2. A diabetic patient on insulin has a plasma phosphate of 0.60 mmol/L (1.9 mg/dL) and a decreased phosphate excretion of 70 mg/24 h. What is the most likely explanation for these results?

a. hyperparathyroidism
b. hypoparathyroidism
c. excess administration of insulin
d. vitamin D deficiency
e. dietary deficiency of phosphorous

Case 3. An 83-year-old man with a history of hypertension and gout was diagnosed 5 years ago with lymphoma and is being seen by a nephrologist. Laboratory results showed the following abnormal results (with reference intervals):

Phosphate: 0.3 mmol/L (0.81−1.45)
Mg: 0.52 mmol/L (0.70−1.23)
LDH: 1312 U/L (110−220)

He was eating well, had no diarrhea, and was not taking any insulin or phosphate binders. His fractional excretion of PO_4 was very low at 1%. This ruled out significant renal loss of PO_4. Because it is a common cause of hypophosphatemia, cellular uptake of PO_4 was considered. Consider the cause of this cellular uptake of PO_4. Hint: it is not a usual cause.

Self-assessment and mastery discussion

Assessment of Case 1. This patient has pancreatitis, consistent with many of his laboratory results. These include the elevated amylase and lipase, the respiratory alkalosis, and the decreased calcium levels. The massive inflammatory response of pancreatitis, similar to sepsis, can release many factors, such as cytokines, that can cause electrolyte abnormalities, including hypophosphatemia.

Assessment of Case 2. Answer c. As the administration of insulin causes a shift of phosphate from extracellular to intracellular compartments, the most likely cause of the decreased plasma and urine phosphorous in this diabetic patient, would be excess administration of insulin. Hyperparathyroidism can lower plasma phosphorous but should increase urinary phosphorous. A chronic dietary deficiency of phosphorous could also cause these findings, but there is no evidence for this situation occurring.

Assessment of Case 3. While insulin, alkalosis, and refeeding after starvation can cause rapid movement of PO_4 into cells, a less common cause of cellular uptake is from rapidly proliferating cells, as with this patient's lymphoma. Giving IV PO_4 solution only slightly increased his PO_4 concentration. However, when high-dose corticosteroids were given, the patient's symptoms resolved within 3 days.

This appeared to be a case of tumor genesis syndrome, in which the rapidly proliferating cells take up PO_4 so rapidly that the normal homeostatic regulation cannot maintain normal serum PO_4 concentrations *(15)*.

References

1. Lederer, E.; Batuman, V. *Hypophosphatemia;* eMedicine; Medscape, 2020. https://emedicine.medscape.com/article/242280.
2. Levine, B. S.; Kleeman, C. R. Hypophosphatemia and Hyperphosphatemia: Clinical and Pathophysiologic Aspects. In *Maxwell and Kleeman's Clinical Disorders of Fluid and Electrolyte Metabolism;* Narins, R. G., Ed., 5th ed.; McGraw-Hill: New York, 1994; pp 1045−1097.
3. Wadsworth, R. L.; Siddiqui, S. Phosphate Homeostasis in Critical Care. *BJA Edu.* **2016,** *16* (9), 305−309.
4. Weiss-Guillet, E.-M.; Takala, J.; Jakob, S. M. Diagnosis and Management of Electrolyte Emergencies. *Best Pract. Res. Clin. Endocrinol. Metab.* **2003,** *17,* 623−651.
5. Blaine, J.; Chonchol, M.; Levi, M. Renal Control of Calcium, Phosphate, and Magnesium Homeostasis. *Clin. J. Am. Soc. Nephrol.* **2015,** *10,* 1257−1272.
6. Toffaletti, J. G. Calcium, Magnesium, and Phosphate. In *Clinical Laboratory Medicine;* McClatchey, K. D., Ed.; Williams and Wilkins: Baltimore, 1994; pp 387−401.
7. Barak, V.; Schwartz, A.; Kalickman, I.; Nisman, B.; Gurman, G.; Shoenfeld, Y. Prevalence of Hypophosphatemia in Sepsis and Infection: The Role of Cytokines. *Am. J. Med.* **1998,** *104,* 40−47.
8. Felsenfeld, A. J.; Levine, B. S. Approach to Treatment of Hypophosphatemia. *Am. J. Kidney Dis.* **2012,** *60* (4), 655−661.
9. Stubbs, J. R.; Yu, A. S. *Overview of the Causes and Treatment of Hyperphosphatemia.* UpToDate, June 2020. https://www.uptodate.com/contents/overview-of-the-causes-and-treatment-of-hyperphosphatemia.
10. Lederer, E.; Batuman, V. *Hyperphosphatemia;* eMedicine; Medscape, December 2020. https://emedicine.medscape.com/article/241185.
11. Adam, M.; Ardinger, H.H.; Pagon, R. A.; et al. *Age-Based Normal Serum Phosphate Reference Intervals.* University of Washington. http://www.genereviews.org, copyright 1993−2021.
12. Sofronescu, A. G.; Staros, E. B. *Phosphate (Phosphorus);* eMedicine.Medscape, November 2019. https://emedicine.medscape.com/article/2090666.
13. *Phosphorus, Urine.* ARUP Lab Test Directory. https://ltd.aruplab.com/Tests/Pub/0020478 (accessed March 2021).
14. Bellasi, A.; Di Micco, L.; Russo, D.; De Simone, E.; Di Iorio, M.; Vigilante, R.; et al. Fractional Excretion of Phosphate (FeP) Is Associated with End-Stage Renal Disease Patients with CKD 3b and 5. *J. Clin. Med.* **2019,** *8,* 1026. https://doi.org/10.3390/jcm8071026.
15. Radi, S. A.; Nessim, S. J. Case: Severe Hypophosphatemia in a Patient with Lymphoma. *Kidney Int.* **2020,** *98,* 243−244.

Chapter 8

Osmolality, sodium, potassium, chloride, and bicarbonate

Osmolality and volume regulation

Introduction

Osmolality in plasma is related to the number of solutes (soluble particles) dissolved in a kg of plasma water, with a normal osmolality between 280 and 295 mOsm/kg. Measurement of serum osmolality is used to evaluate the body's regulation of water and sodium balance, while urine osmolality evaluates the kidney's ability to concentrate urine. In a steady state, our total body water (TBW) and salt content remain relatively constant. An increase or decrease in water and salt intake will affect the water and salt retention and excretion in the kidney.

The TBW is defined as the fluid that occupies the intracellular ($\sim$65% of TBW) and extracellular spaces ($\sim$10% intravascular; $\sim$25% interstitial) of the body. In general, while TBW fluctuates hourly due to loss via the kidneys, lungs, and skin, and gain by food and fluid intake, plasma osmolality is more closely regulated and remains relatively constant *(1,2)*.

Note that the term "osmolarity" is related to the number of solutes per *liter* of plasma volume and is expressed in mmol/L. Another term is the oncotic or colloid osmotic pressure, which is related to the water-retaining effect of proteins. Clinically, this is most relevant in the intravascular fluid (blood) where the high protein concentration in blood plasma prevents excess water movement to the interstitial fluid. Disturbance to this equilibrium can cause edema, which is an accumulation of fluid in the interstitial space.

Water excretion in the kidneys is largely controlled by antidiuretic hormone (ADH) also called vasopressin, which is a peptide secreted by the posterior pituitary. To increase water reabsorption, ADH opens water channels in the membranes of cells lining the collecting ducts in the kidney *(3)*.

Several factors affect ADH secretion: (1) receptors in the hypothalamus stimulate ADH production and secretion in the posterior pituitary in response to increasing osmolality; (2) Stretch receptors located in the atria of the heart sense increased blood pressure due to increased blood volume and inhibit

Blood Gases and Critical Care Testing. https://doi.org/10.1016/B978-0-323-89971-0.00004-5
159

ADH secretion; and (3) stretch receptors in the arteries stimulate ADH secretion when blood pressure falls due to decreased blood volume.

The sodium concentration in blood is affected by the regulation of both the plasma osmolality and the blood volume. As we will see later, osmolality and volume are regulated by separate mechanisms, although both mechanisms involve ADH: osmolality (sodium) is regulated by changes in water balance via ADH, whereas volume is regulated by changes in sodium balance via the renin—angiotensin—aldosterone system and ADH *(4,5)*. In a person with normal water and osmolality homeostasis, if a large volume of hypotonic fluid such as water is consumed, to prevent volume overload the kidneys respond by rapidly excreting dilute urine, even before the intra- and extracellular fluids equilibrate *(2)*.

Calculation of osmolality

The osmolality of plasma is related to the concentration of ions (Na, K, Cl, albumin, etc.) and neutral solutes such as glucose and urea. The contribution of the ions can be fortuitously estimated as $2 \times$ [Na] (in mmol/L), for three reasons:

- NaCl in plasma is 75% "osmotically active." Therefore, 1 mmol of NaCl behaves as if it dissociates into 1.75 mmol of osmotic particles (0.75 Na^+ + 0.75 Cl^- + 0.25 NaCl) *(6)*.
- Plasma is 93% water and 7% proteins and lipids (with ions mostly confined to the plasma H_2O space).
- Other ions such as K^+, Ca^{2+}, and Mg^{2+} account for $\sim 8\%$ of the osmotic activity relative to sodium ion.

Therefore, sodium contribution to osmolality is

$$(1.75 / 0.93) \times 1.08 \times [\text{Na}] = 2 \times [\text{Na}]$$

Calculated osmolality (mOsm/kg) =

$$(2 \times [\text{Na}]) + [\text{BUN}(\text{mg} / \text{dL}) / 2.8] + [\text{glucose}(\text{mg} / \text{dL}) / 18]$$

where BUN is blood urea nitrogen.

Note: Because concentrations of sodium, urea, and glucose are in molarity, calculations of "osmolality" are actually calculations of osmolarity.

Regulation of osmolality

The response to increased osmolality is described in the following and shown in Fig. 8.1.

Renal water regulation by ADH and thirst each play important roles in regulating plasma osmolality. To maintain a normal plasma osmolality

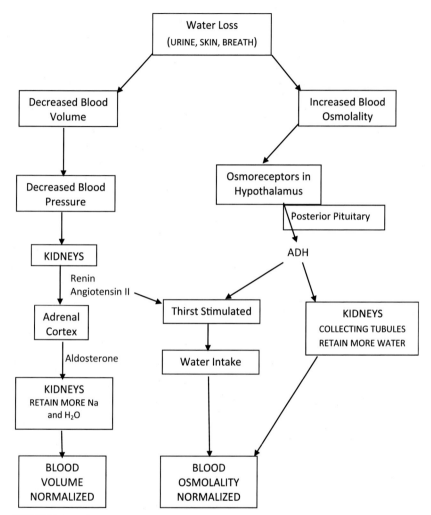

FIGURE 8.1   Responses to changes in blood volume and osmolality. Loss of water causes both hypovolemia and increased blood osmolality. Hypovolemia decreases blood pressure, which stimulates both the renin—angiotensin—aldosterone system and secretion of ADH, with ADH being a common response to both hypovolemia and hyperosmolality. Volume receptors located in the afferent renal arterioles, the heart, the arteries, and the distal tubule, by sensing decreased blood pressure, stimulate the secretion of renin. Increased blood osmolality is detected by osmoreceptors in the hypothalamus, which produces ADH that is stored and released through the posterior pituitary gland. ADH has several actions that normalize blood osmolality. It activates the kidney tubules of the collecting duct to reabsorb more water by producing water channel proteins called aquaporins. ADH causes vasoconstriction to temporarily increase blood pressure. ADH also leads to increased thirst that should cause increased water intake to normalize blood osmolality. *This is an original drawing.*

[$\sim$280–300 mmol/kg ($\sim$280–300 mOsm/kg) of plasma H_2O)], both thirst and ADH are activated. Osmoreceptors in the hypothalamus respond quickly to small changes in osmolality: A 1%–2% increase in osmolality causes a fourfold increase in the circulating concentration of ADH and a strong desire to drink fluid. Conversely, a 1%–2% decrease in osmolality shuts off ADH production entirely (7). ADH acts by increasing the reabsorption of water in the cortical and medullary collecting tubules of the kidney. The action is short-lived because ADH has a half-life in the circulation of only 15–20 min. The adrenal cortex also affects aldosterone secretion by shutting off secretion when plasma osmolality increases (3).

Renal water excretion is more important in controlling water excess, whereas thirst is more important in preventing water deficit or dehydration. Consider what happens in the following conditions:

Water Load. Excess intake of water (either done normally or in a condition such as polydipsia) is handled by several mechanisms. As plasma osmolality declines, both ADH and thirst are suppressed. In the absence of ADH, a very large volume of dilute urine can be excreted (10 L or more per day), well above any normal intake of water. If blood volume increases, stretch receptors located along the circulatory system stimulate the production of natriuretic peptides, including atrial (ANP) and "brain" (BNP) natriuretic peptides. These mechanisms are so effective that hypo-osmolality and hyponatremia occur almost exclusively in patients with excess ADH and/or impaired renal excretion of water (6).

Water Deficit. As a deficit of water begins to increase plasma osmolality, both ADH secretion and thirst are activated. Even though ADH minimizes renal water loss, thirst is the major defense against hyperosmolality and hypernatremia because it stimulates a person to seek an external source of water. As an example of the effectiveness of thirst in preventing dehydration, a patient with diabetes insipidus (no ADH) may excrete 10 L of urine per day. However, because thirst persists, water intake matches output to maintain a normal plasma sodium concentration (6).

This is why hypernatremia rarely occurs in a person with a normal thirst mechanism and access to water. However, it becomes a concern in infants, unconscious patients, or anyone who is unable to either drink or ask for water. In people over the age of 60 years, osmotic stimulation of thirst diminishes. Particularly in the older patient with illness and diminished mental status, dehydration becomes increasingly likely.

Regulation of blood volume

The response to decreased blood volume is described in the following and shown in Fig. 8.1.

Blood volume is the total amount of fluid within the arteries, capillaries, veins, and chambers of the heart. The blood volume also includes red blood cells (erythrocytes), white blood cells (leukocytes), platelets, and plasma. Plasma accounts for about 60% of total blood volume while erythrocytes and the other cells make up roughly 40% *(5)*.

Adequate blood volume maintains blood pressure and ensures good perfusion to tissues and organs. Multiple organ systems are involved in regulating blood volume by the interrelated control of both sodium and water. The kidneys are primarily responsible for regulating blood volume by adjusting the solute and water content of the blood through filtration, reabsorption, and secretion.

Changes in blood volume are initially detected as changes in blood pressure by a series of stretch receptors in areas such as the cardiopulmonary circulation, the carotid sinus, the aortic arch, and the glomerular arterioles. In the kidney nephrons, Na^+ and K^+ are actively reabsorbed in the proximal tubules, with Cl^- and water "following" the reabsorption of the cations. The amounts of water and solute reabsorbed primarily regulate blood volume. If blood volume is too low, more filtrate is reabsorbed; if blood volume is too high, less filtrate is reabsorbed *(5)*.

The renin-angiotensin-aldosterone system responds primarily to approximately a 5% decrease in blood volume. While a potent thirst response is also stimulated, it requires a 5%−10% decrease in blood volume and arterial pressure. While this response to decreased blood volume is less sensitive than the thirst response to only a 1%−2% decrease in plasma osmolality, once activated, this response to decreased volume dominates the response to changes in osmolality *(7)*.

The effects of blood volume and osmolality on sodium and water metabolism are listed in the following and also shown in Fig. 8.1.

- In response to decreased blood volume and blood pressure, the kidneys respond by decreasing GFR and secreting renin near the renal glomeruli (juxtaglomerular cells). Renin converts angiotensinogen to angiotensin I, which then becomes angiotensin II.
- Angiotensin II stimulates thirst, causes vasoconstriction to quickly increases blood pressure, stimulates the adrenal cortex to secrete aldosterone, and increases renal reabsorption of Na.
- Aldosterone promotes distal tubular reabsorption of Na^+ and HCO_3^- in exchange for K^+ and H^+.
- In volume depletion, aldosterone increases blood volume by increasing renal retention of sodium and bicarbonate, and the water that accompanies these ions. In this process, both Cl^- and H^+ are lost *(8)*.
- Both thirst and ADH are stimulated by low blood volume, independent of osmolality.
- Natriuretic peptides (ANP and the now-famous BNP) are released from the myocardium in response to excess blood volume and promote sodium excretion in the kidney.

- Epinephrine and norepinephrine are secreted in response to decreased blood volume.
- All other things being equal, an increased plasma sodium will increase urinary sodium excretion, and vice versa.

The renal retention of sodium has a profound effect on blood volume because whenever a sodium ion is reabsorbed, a water molecule follows. While large amounts of sodium are filtered in the 150 L of glomerular filtrate produced daily, the renal tubules reabsorb 98%−99% of this sodium along with most of the water. Thus, even a 1%−2% reduction in tubular reabsorption of Na^+ can increase water loss by several liters per day.

Urine osmolality values may vary widely depending on water intake and the circumstances of collection. However, it is generally decreased in diabetes insipidus (inadequate ADH) and polydipsia (excess H_2O intake due to chronic thirst) and increased in conditions where ADH secretion is increased by hypovolemia or hyperosmolality or in conditions such as inappropriate ADH.

Reference interval

The reference intervals for osmolality, electrolytes, and other parameters are shown in Table 8.1.

Sodium

Physiology of sodium balance

As an electrolyte, sodium is vital for the transmission of nerve impulses and activation of muscle movements. Neurons generate electrical signals through brief, controlled changes in the permeability of their cell membrane to ions such as Na^+ and K^+ (9). Sodium is the most abundant cation in the ECF, representing 90% of all extracellular cations, and largely determines the osmolality of the ECF. To maintain the much higher sodium concentration in the ECF relative to the intracellular concentration, an active transport mechanism involving a Na-K-ATPase pump maintains this large gradient between ECF and ICF. This Na-K-ATPase pump exchanges three sodium ions moving out of cells for two potassium ions moving into cells (4).

The plasma sodium concentration depends greatly on the intake and excretion of water and, to a somewhat lesser degree, the renal regulation of sodium. Three processes are of primary importance:

- The intake of water in response to thirst, as stimulated or suppressed by plasma osmolality
- The excretion of water, largely affected by ADH release in response to changes in either blood volume or osmolality
- The blood volume status, which affects sodium excretion through aldosterone, angiotensin II, and ANP

TABLE 8.1 Reference intervals for serum/plasma electrolytes and osmolality[a].

	SI units	Conventional units
Sodium		
(2 years–adult)	135–145 mmol/L	135–145 mEq/L
Potassium		
0–30 days	4.0–6.0 mmol/L	4.0–6.0 mEq/L
1 month–2 years	4.0–5.5 mmol/L	4.0–5.5 mEq/L
2–18 years	3.8–5.2 mmol/L	3.8–5.2 mEq/L
18 years–adult	3.5–5.0 mmol/L	3.5–5.0 mEq/L
Chloride		
0–2 years	96–110 mmol/L	96–110 mEq/L
2 years–adult	98–108 mmol/L	98–108 mEq/L
Total CO_2		
0–17 years	18–27 mmol/L	18–27 mEq/L
18 years–adult	21–30 mmol/L	21–30 mEq/L
Anion gap		
Na, Cl, HCO_3	4–12 mmol/L	4–12 mEq/L
Na, K, Cl, HCO_3	8–16 mmol/L	8–16 mEq/L
Osmolality, plasma		
Adult	280–300 mmol/kg	280–300 mOsm/kg
Osmolality, urine		
(24-h collection)	300–900 mmol/kg	300–900 mOsm/kg

[a]All values are plasma values unless noted otherwise.

As people age, they are less able to maintain fluid and sodium balance for several reasons. Their thirst sensation diminishes so they may not drink fluids when needed. Kidney function diminishes, such that kidneys may become less able to regulate water and electrolytes in the blood. Also with age, the body contains less fluid, which means that a slight loss of fluid and sodium, as from a fever or from lost thirst or appetite, can be more serious.

People who have physical or mental illnesses may not drink fluids even when they are thirsty. Commonly used drugs for high blood pressure, diabetes mellitus, or heart disorders can affect fluid excretion and magnify the effects of fluid loss *(10)*.

The kidneys have the ability to conserve or excrete large amounts of sodium, depending on the osmolality (mostly sodium) of the ECF and the blood volume. Normally, the kidney reabsorbs 98%−99% of filtered sodium: 60%−75% is reabsorbed in the proximal tubule, with the remainder reabsorbed along with Cl^- in the loop and distal tubule and exchanged for K^+ in the connecting segment and cortical collecting tubule. These ion movements are under the control of aldosterone, which regulates the final handling of Na ions in the distal nephron by its control of a Na^+-Cl^- cotransporter (NCC) and an Epithelial Na^+ Channel, both of which promote reabsorption of Na ion in the distal tubule *(11)*.

Frequency of hyponatremia and hypernatremia

Mild hyponatremia is quite common among hospitalized patients, with 15%−20% having a serum sodium level of <135 mmol/L, but only 1%−4% have a serum sodium level of less than 130 mmol/L *(12)*. A large study of hospitalized patients found that even mild degrees of hyponatremia were associated with increased 1-year and 5-year mortality rates. Mortality was particularly increased in those with cardiovascular disease, metastatic cancer, and those undergoing orthopedic procedures *(13)*.

Hypernatremia on admission to the hospital has an estimated incidence from about 2% to 25% *(14)*. A seemingly more appropriate incidence found that among patients admitted to the ICU, 11% developed mild hypernatremia and 4.2% developed moderate-to-severe hypernatremia within 24 h after admission *(15)*.

Hyponatremia

Causes of hyponatremia

Hyponatremia is defined as a plasma sodium concentration <135 mmol/L, with its duration also an important consideration in treatment *(12)*. Symptoms include anorexia, nausea, confusion, lethargy, easy agitation, and headache. At values below 120 mmol/L, general weakness and mental confusion are typical. Paralysis and severe mental impairment may occur at plasma sodium <115 mmol/L *(16)*. Guidelines that classify hyponatremia based on the plasma sodium concentration have been published in Europe *(17)*:

Mild: 130−134 mmol/L
Moderate: 125−129 mmol/L
Profound: <125 mmol/L

In hyponatremia, water moves into cells, causing them to swell. If the onset of hyponatremia is longer than 48 h, treatment must be gradual to avoid osmotic demyelination from overly rapid correction. This is especially dangerous for brain cells because expansion increases intracranial pressure. Cerebral edema may lead to respiratory insufficiency and hypoxia, which can cause death *(12,18)*.

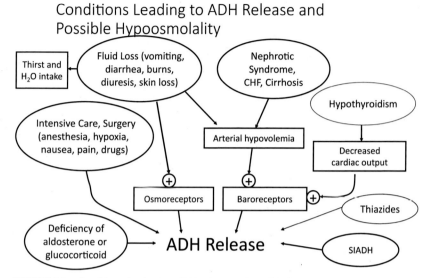

FIGURE 8.2 Conditions that lead to ADH release and possible hyponatremia. *CHF*, congestive heart failure; *SIADH*, syndrome of inappropriate secretion of antidiuretic hormone. *From Fig. 3.2 in 2nd ed.*

Most patients with hyponatremia have an excess of ADH, either from an inappropriate secretion of ADH or in response to a decreased blood volume *(19)*, as shown in Fig. 8.2. History is important in evaluating patients with hyponatremia for conditions such as heart failure, cirrhosis, GI losses, burns, endocrine disorders, excessive sweating, or certain drugs. Chronic hyponatremia is a common clinical problem in the elderly.

Laboratory evaluation of hyponatremia

Detecting, monitoring, and treating abnormal electrolyte concentrations are critically important in many patients, with rapid results often reducing therapeutic intervention time *(20)*. The most important laboratory tests for evaluating patients with hyponatremia are:

- Plasma Na, K, Cl, HCO_3^-, glucose, creatinine, and urea.
- Serum osmolality
- Urine osmolality and Na

The diagnostic approach to hyponatremia is shown in Table 8.2. The initial decreased plasma Na should be confirmed by a decreased serum osmolality.

Serum osmolality confirms true hyponatremia versus pseudohyponatremia due to hyperlipidemia or hyperproteinemia, or hypertonic hyponatremia from elevated glucose, urea, or administration of some osmotic substance such as mannitol.

TABLE 8.2 Evaluation of hyponatremia.

Measure plasma Na, K, Cl, HCO_3^-, and urea; osmolality and uric acid if needed.

Plasma Na and osmolality decreased:

Measure urine Na:

    Urine Na <15 mmol/L (<15 mEq/L); plasma AG, urea, and uric acid normal to increased.

        Hypovolemia with hypotonic fluid replacement (diarrhea, vomiting, sweating, renal losses, etc.)

        Polydipsia (chronic thirst)

        Hypervolemia with arterial hypovolemia (congestive heart failure, cirrhosis)

    Urine Na >20 mmol/L (>20 mEq/L); plasma AG, urea, and uric acid normal to decreased.

        Inappropriate excess secretion of ADH (carcinoma of lung, adrenal insufficiency, etc.)

        Renal salt loss (thiazides, aldosterone deficiency)

        Reset osmostat

        Renal failure with water overload

The urine osmolality and sodium help differentiate between renal and neurogenic causes of hyponatremia, such as polydipsia. If the urine osmolality is greater than about 150 mOsm/kg, it indicates the impaired ability of the kidneys to produce dilute urine.

Clinical diagnosis of hyponatremia

A clinical history, symptoms, and physical exam are required along with the initial laboratory parameters that indicated hyponatremia. Symptoms include anorexia, nausea, confusion, lethargy, easy agitation, and headache *(21)*. The history should evaluate the following:

- Duration of the hyponatremia, with 48 h a typical cutoff between acute and chronic hyponatremia.
- Edema (heart failure, cirrhosis)
- GI losses (diarrhea, vomiting)
- Skin losses (burns, sweating)
- Renal losses (low aldosterone, diuretics, salt-losing syndromes)
- Drugs (carbamazepine, SSRIs)
- Cancer (oat-cell carcinoma of the lung is associated with SIADH)
- Check blood pressure and examine skin:
  Volume depletion: decreased BP; skin has decreased turgor, and is cool and pale
  Volume overload: elevated BP, pitting edema in lower extremities

TABLE 8.3 Hyponatremia differential diagnosis related to volume status.

Hypovolemic	Na loss in excess of H_2O
	Thiazide diuretics Loss of hypertonic fluid: GI, burns, sweat Potassium depletion Aldosterone deficiency Salt-losing nephropathies
Euvolemic	Problem with water balance
	Excess or inappropriate ADH secretion Artifactual—severe hyperlipidemia Hyperosmolar from substance other than sodium Polydipsia Adrenal insufficiency Altered regulatory set point for osmolality Drugs
Hypervolemic	Movement of fluid from intravascular to interstitial space
	Congestive heart failure, hepatic cirrhosis Advanced renal failure (decreased glomerular filtration rate) with excess water intake Nephrotic syndrome—decreased colloid osmotic pressure

The pathogenesis of hyponatremia is often related to volume status and excess ADH that is secreted in response to changes in volume or osmotic stimuli, as shown in Table 8.3 and Fig. 8.2. Volume status may be assessed by skin turgor, jugular venous pressure, and urine Na concentration, with a low urine Na indicating hypovolemia (1,3,4).

Hyponatremia is classified according to volume status, as follows:

Hypovolemic hyponatremia: a decrease in intravascular (plasma) water with a greater decrease in plasma sodium. Among the more common causes of hypovolemic hyponatremia are the following:

- Use of thiazide diuretics (but not loop diuretics), which induce Na and K loss without interfering with ADH-mediated water retention.
- With prolonged vomiting, diarrhea, or sweating, as hypotonic fluids are lost they are replaced by a relatively greater volume of hypotonic fluid by ingestion of water in response to thirst and ADH, which are stimulated by hypovolemia (see Fig. 8.2).
- In potassium depletion, cellular loss of K^+ promotes Na^+ movement into the cell with an associated decrease in plasma Na and volume.
- Aldosterone deficiency from adrenal insufficiency or drugs increases both urinary Na loss and urine osmolality.
- Salt-wasting nephropathy may infrequently develop in renal tubular and interstitial diseases, such as medullary cystic and polycystic kidney

diseases, usually as renal insufficiency becomes severe [serum creatinine >610 μmol/L (>8 mg/dL)].

Normovolemic hyponatremia: little or no increase in plasma water with some loss of plasma sodium. It typically indicates a problem with water balance and may be related to one of the following:

- Polydipsia (chronic thirst with excess intake and excess excretion of water) eventually leads to hyponatremia that is usually mild but occasionally severe.
- Excess ADH may be secreted in response to drugs, surgery, tumors, central nervous system disorders, endocrine disorders, or pulmonary conditions *(4,19)*. The excess ADH causes mild hypervolemia, which then leads to excretion of Na^+ and water by the release of ANP (see Fig. 8.2).
- Pseudohyponatremia can occur with severe hyperlipidemia or hyperproteinemia. Methods that dilute plasma before Na analysis by the ion-selective electrode (ISE) or other methods give erroneously low Na results on such samples by measuring the Na concentration per liter of plasma. "Direct" methods by ISE (typically on blood gas analyzers), which do not dilute plasma or whole blood, give accurate Na results because they detect the Na concentration in the plasma water.
- A significant degree of hyperglycemia generally causes a lower plasma sodium to maintain a normal plasma osmolality.
- In adrenal insufficiency, the decreased aldosterone (mineralocorticoid) and cortisol (glucocorticoid) promote ADH release *(3)*.
- In pregnancy, osmotic regulation in the hypothalamus may be offset such that plasma Na is regulated ∼5 mmol/L (∼5 mEq/L) lower than normal. This effect on the hypothalamus may be initiated by vasodilation, which mimics hypovolemia by lowering blood pressure.
- Drugs: desmopressin, psychoactive agents, anticancer agents *(22)*.

Hypervolemic hyponatremia: total body sodium is increased, but with an even greater increase in TBW. This condition is nearly always a problem of water overload as excess water accumulates from renal failure, heart failure, or liver failure. It is usually associated with one or more of the following conditions:

- Excess secretion of ADH. For example, congestive heart failure or hepatic cirrhosis increases venous back-pressure in the circulation, which promotes the movement of fluid from the blood to the interstitium, causing edema and arterial hypovolemia. The arterial hypovolemia stimulates ADH secretion, which eventually leads to hypervolemia and hyponatremia, as indicated in Fig. 8.2.
- Inability of the kidneys to excrete excess water along with excess fluid intake. This is more likely in elderly patients.

- Aldosterone secretion stimulates renal Na ion and Cl ion reabsorption in the distal convoluted tubule (DCT) by controlling both a sodium-chloride cotransporter (NCC) and an epithelial Na channel (ENaC) in the DCT. Aldosterone also regulates a sodium channel in the collecting duct and the chloride-reabsorbing protein pendrin in the cortical collecting duct (23).

Treatment of hyponatremia

The treatment of hyponatremia depends on the cause, the severity, and how rapidly the hyponatremia develops (12,13,21,22,24). In patients with good kidney function and less severe symptoms, fluid restriction is usually effective (21).

Acute hyponatremia is less common than chronic hyponatremia and is typically seen in patients with a sudden ingestion of free water. The danger for such patients is herniation of the brainstem if sodium concentrations fall below 120 mmol/L (21). The therapeutic goal in acute hyponatremia is to increase the serum sodium level rapidly by 4−6 mmol/L over the first 1−2 h. In addition, the source of free water must be identified and eliminated. In patients with healthy renal function and without severe symptoms, the plasma sodium may correct without further intervention. However, patients with seizures, severe confusion, coma, or signs of brainstem herniation should receive hypertonic (3%) saline to rapidly correct the serum sodium concentration, but only enough to alleviate the symptoms (21).

Patients with chronic hyponatremia and more severe symptoms, such as severe confusion, coma, or seizures, should receive hypertonic saline, but only enough to raise the serum sodium level by 4−6 mmol/L and to arrest seizure activity. After this increase, no further correction of the sodium during the next 24 h is recommended (21).

Hypervolemic hyponatremia or asymptomatic hyponatremia is usually treated effectively with water restriction. This would be the case with advanced renal disease, where the inability to excrete water promotes hypervolemia whenever fluid intake is excessive.

Hypernatremia

Causes of hypernatremia

Hypernatremia is defined as an abnormal plasma sodium concentration above 145 mmol/L. Hypernatremia is usually a problem of water regulation more than a problem of sodium homeostasis so that hypernatremia usually results from excessive loss of water relative to sodium (14). Hypotonic fluid may be lost through several common routes: by the kidney, by profuse sweating, or by GI loss such as diarrhea. About 1 L of water per day is normally lost through the skin and by breathing (insensible losses). Any condition that increases water loss, such as fever, burns, or exposure to heat, will increase the likelihood of developing hypernatremia. Because an increase in osmolality of only 1%−2% stimulates thirst, hypernatremia rarely occurs in persons with a normal thirst mechanism and access to water.

Hypernatremia and hyperosmolality from the water loss can cause neuronal cell shrinkage and brain injury, while the loss of volume can lead to circulatory problems, such as tachycardia and hypotension *(14)*. Severe symptoms usually occur with an acute increase in sodium concentration to approximately 158−160 mmol/L *(22)*. Cerebral dehydration can lead to demyelination, cerebral bleeding, coma, and death *(24)*. Because the brain can adapt to the slower (chronic) onset of hypernatremia by restoring normal cell volume, chronic hypernatremia is less likely to induce neurologic symptoms *(22)*.

Acute hypernatremia, which is defined as occurring in a period of less than 24 h, should be corrected rapidly. Chronic hypernatremia occurring over a period of longer than 48 h should be corrected more gradually to avoid cerebral edema during treatment. Hypernatremia should be corrected at a rate no faster than about 0.5 mmol/L/h *(13)*.

Laboratory and clinical evaluation of hypernatremia

Following the detection of hypernatremia as a serum sodium concentration above 145 mmol/L, the following laboratory studies may be helpful in evaluating the cause:

Serum electrolytes (Na^+, K^+, Ca^{++}, Cl^-), glucose, urea, and creatinine
Urine and plasma osmolality; also urine Na^+ and K^+ may be helpful
24-h urine volume
In more challenging cases, measuring plasma ADH may be helpful.

Severe hypernatremia occurs when the plasma sodium concentration reaches ∼158−160 mmol/L, or when symptoms due to hypernatremia are severe. The measurement of urine osmolality is necessary to evaluate the cause of hypernatremia, as shown in Table 8.4. In renal losses of water, the urine osmolality is low or normal. In nonrenal losses, the urine osmolality is increased.

Nonrenal causes of hypernatremia are common, with examples as follows:

- Inadequate fluid intake to replace water that is lost continuously by breathing, sweating, evaporation through skin, or by GI loss of hypotonic fluid.
- Altered mental status in elderly patients and in infants, both of whom may have an intact thirst mechanism, but who are unable to ask for or obtain water.
- Diminished or absent thirst reflex (hypodipsia), which normally diminishes with age.

TABLE 8.4 Evaluation of hypernatremia.[a]

If plasma Na is >150 mmol/L (150 mEq/L), measure urine osmolality.
>800 mmol/kg (>800 mOsm/kg) (inadequate fluid intake)
    Insensible losses of water
    GI loss of hypotonic fluid
    Diminished or lost thirst or inability to obtain water
    Excess i.v. or oral intake of Na
300–800 mmol/kg (300–800 mOsm/kg)
    Impaired ADH release (hyperaldosteronism, Cushing's syndrome)
    Partial central diabetes insipidus
    Diuretics
    Osmotic diuresis
<300 mmol/kg (<300 mOsm/kg)
    Diabetes insipidus (central or nephrogenic)

[a]Persons who cannot fully concentrate their urine, such as neonates, young children, the elderly, and some patients with renal insufficiency, may show a relatively lower urine osmolality.

- Ingestion of too much salt is an obvious but uncommon primary cause of chronic hypernatremia. Similarly, hypernatremia caused by i.v. administration of excess hypertonic saline is uncommon but more likely in neonates.

In the nonrenal examples above, ADH secretion is increased to minimize renal loss of water, resulting in very high urine osmolality: >800 mmol/kg (>800 mOsm/kg), although older persons may be less able to concentrate urine and produce a urine osmolality that high (Table 8.4).

Renal causes of hypernatremia are often related to drugs, such as diuretics, or processes that interfere with the ability of the kidney to concentrate their urine. Insufficient production or unresponsiveness to ADH are general mechanisms for hypernatremia by increasing renal water loss. Many drugs can induce decreased responsiveness to ADH *(14)*. Diabetes insipidus is caused by any pathologic process that destroys the anatomic structures of the hypothalamus or pituitary that produce and secrete ADH *(14)*. Hyperglycemia can elevate plasma osmolality simply by glucose-mediated urinary loss of hypotonic fluids during osmotic diuresis.

Chronic hypernatremia in an alert patient with normal thirst is indicative of hypothalamic disease. This includes the following:

- Primary hyperaldosteronism, in which excess aldosterone enhances reabsorption of sodium and excretion of potassium and hydrogen ion.
- Cushing's syndrome, with excess production of cortisol *(4)*.
- ADH deficiency may be caused by impaired ADH secretion (diabetes insipidus) or by impaired renal response to ADH (nephrogenic diabetes insipidus). Because increased thirst typically compensates for renal water

loss, hypernatremia does not usually occur in diabetes insipidus unless the thirst mechanism is also impaired. Diabetes insipidus is characterized by copious production of dilute urine (3−20 L/d). A partial defect in either ADH release or ADH effect at the receptor level will hinder the ability of the kidneys to sufficiently reabsorb water to correct the hypernatremia.

- Excess water loss may also occur in renal tubular disease, such as acute tubular necrosis, in which the tubules become unable to fully concentrate the urine.

Treatment of hypernatremia

After detecting hypernatremia and recognizing symptoms, the underlying cause must be identified, followed sequentially by correction of volume disturbances and correction of the hypernatremia. Correcting hypernatremia by replacing water loss must be done carefully with the rate of sodium correction depending on how acutely the hypernatremia developed and on the severity of symptoms.

In patients who develop symptomatic hypernatremia acutely over a period of less than 24 h, rapid correction of the hypernatremia is beneficial with little risk of cerebral edema *(22)*. Typically, acute hypernatremia can be corrected at about 1 mmol/L/h.

With chronic hypernatremia occurring over a period longer than 48 h, correction should be more gradual due to the risks of brain edema during treatment. Because the brain adjusts to and mitigates chronic hypernatremia by increasing the intracellular osmolality, if serious hypernatremia (>160 mmol/L) is corrected too rapidly, water movement into the brain cells can cause cerebral edema and may lead to herniation and permanent neurologic deficits *(14)*. Chronic hypernatremia should be corrected at a maximal rate of 0.5 mmol/L/h until the plasma Na reaches 145 mmol/L *(14,22)*.

Reference intervals for sodium are shown in Table 8.1.

Potassium

Physiology

Potassium is the major intracellular cation, with about 98% in the cells and only about 2% in the blood *(25)*, which is maintained within a narrow concentration from approximately 3.5 to 5.1 mmol/L. Cellular membrane potentials and other functions require that the body maintain an appropriate ratio of K^+ concentrations between the cell and ECF *(9)*. Insulin, B-adrenergic catecholamines, thyroid hormone, and acid−base balance influence this cellular transport.

Potassium has several vital functions *(26)*, including the following:

- Regulation of neuromuscular activity
- Cardiac contraction and cardiac rhythm
- Regulation of intracellular and extracellular volume and acid—base status

The potassium ion concentration in plasma has a major effect on the contraction of cardiac muscle. An increased plasma potassium slows the heart rate by decreasing the resting membrane potential of the cell relative to the threshold potential. A decrease in the plasma potassium concentration to less than 3.5 mmol/L increases the likelihood of arrhythmias. More severe hypokalemia causes general muscle weakness and more severe arrhythmias and atrial tachycardia. At potassium levels of 2.5 mmol/L, myopathy may progress to rhabdomyolysis with myoglobinuria, impairing both kidney function and respiratory function *(27)*.

The K^+ concentration also affects the H^+ concentration in the blood. When hypokalemia is present, K ions are lost from the cells; as a result, Na^+ and H^+ move into the cell to replace K^+. The H^+ concentration is therefore decreased in the ECF, resulting in alkalemia *(4,27)*. The reverse is true for hyperkalemia.

Regulation

Potassium abnormalities may be caused by imbalances in the intake, excretion, or intracellular—extracellular shifts of K^+. The kidneys are primarily responsible for the regulation of potassium balance. Potassium ions are freely filtered by the glomeruli, then the proximal tubules and ascending limb reabsorb nearly all the potassium. The maintenance of potassium homeostasis is controlled by a complex interplay of aldosterone-mediated Na-K exchange mechanisms in the distal tubules and collecting ducts, including NCC (Na/Cl cotransporter), ENaC (Epithelial Na Channel), ROMK (Renal Outer Medullary K) channel, and Na/K-ATPases. For example, a decreased plasma potassium concentration activates NCC activity which reduces K ion excretion. Thus, the distal nephron is the principal determinant of urinary potassium excretion and homeostasis in the blood *(25)*.

Cellular shifts of K^+ are important regulators of plasma K^+ concentrations. An acute oral or i.v. intake of potassium is handled by K^+ uptake from the ECF into the cells. Dietary intake of potassium usually causes the release of insulin that will shift dietary K^+ into cells to normalize plasma K^+. Catecholamines also increase cellular uptake of K^+ to minimize increases in plasma K^+. As the cellular K^+ then gradually returns to the plasma, it is removed by urinary excretion. Because healthy kidneys will excrete potassium even with potassium-deficient diets, prolonged dietary deficiency of potassium can lead to hypokalemia and also exacerbate hypokalemia from other causes

(25). Note that chronic loss of cellular K^+ can result in cellular depletion without appreciable short-term change in the plasma K^+ concentration because any excess K^+ is readily excreted in the urine.

There are several important factors that influence the cellular shifts of K^+ between cells and ECF:

- K^+ loss frequently occurs whenever the Na-K-ATPase pump is inhibited by conditions such as hypoxia, hypomagnesemia, or digoxin overdose.
- Insulin promotes rapid entry of K^+ into skeletal muscle and liver by increasing Na-K-ATPase activity.
- Catecholamines are also factors; for example, β_2-stimulators, such as epinephrine, promote cellular entry of K^+, whereas β-blockers, such as propranolol, impair cellular entry of K^+.
- Alkalemia from either respiratory or metabolic causes promotes K^+ entry into cells. For each 0.1 pH unit increase, plasma K decrease by about 0.4 mmol/L *(25)*. As cells release H^+ to normalize alkalemia, extracellular K^+ and Na^+ enter the cell to maintain electroneutrality. Hypokalemia can also induce alkalosis by increasing reabsorption of HCO_3^- and increasing secretion of acid in the distal nephron *(4)*.

Hypokalemia

Causes of hypokalemia

Causes of hypokalemia are listed in Table 8.5.

Hypokalemia is defined as a serum potassium concentration less than 3.5 mmol/L (3.5 mEq/L. Severity is categorized as mild when the serum potassium level is 3—3.4 mmol/L, moderate when the serum potassium level is 2.5—3 mmol/L, and severe when the serum potassium level is less than 2.5 mmol/L *(26)*. Preanalytical variables can cause slight differences in potassium results between plasma and serum, and significantly increase results if hemolysis is present (see Chapter 10 on Collection and Handling of Samples).

Major causes of hypokalemia are as follows *(26)*:

- Urinary loss of potassium due to diuretics, Mg deficiency, or increased aldosterone or renin.
- GI losses from diarrhea or vomiting.
- Increased cellular uptake of potassium from alkalemia, insulin or glucose administration, aldosterone, catecholamines, caffeine.
- Chronically reduced dietary intake, which is rare as the sole cause.

Renal or GI losses. Renal loss of potassium can result from kidney disorders such as renal tubular acidosis, potassium-losing nephritis, and diabetic ketoacidosis *(25)*. Excess aldosterone can lead to hypokalemia and metabolic alkalosis *(4,6,25)*. Hypomagnesemia can lead to hypokalemia by promoting

TABLE 8.5 Causes of hypokalemia.

Cellular redistribution	
Metabolic alkalosis	Causes cellular uptake of K ions. May be caused by vomiting.
Insulin excess	From administration or postprandial release of insulin. Promotes movement of K ions into cells
B-adrenergic agonists	Stimulate Na/K-ATPase pumps
Drugs, such as epinephrine and caffeine	Epinephrine stimulates B-2-receptors causing intracellular shift of potassium. Caffeine increases urinary loss of K by its diuretic action
Increased kidney excretion	
Mineralocorticoid excess	Aldosterone increases secretion of K ions in the nephron tubules to increase urinary loss.
Diuretics	Thiazide diuretics increase renal K loss
Mg deficiency; hypomagnesemia	Mg deficiency promotes renal loss of K ion. Mg must be normalized to treat hypokalemia.
GI loss	
Diarrhea or use of laxatives	Less K is absorbed in the intestines.
Vomiting	Gastric fluid contains modest K, but vomiting can cause volume depletion and metabolic alkalosis, as noted earlier.
Dietary deficiency	
Chronic low K diet	Usually occurs with diminished kidney function

both urinary and fecal loss of potassium. Magnesium deficiency diminishes the activity of Na-K-ATPase and enhances the secretion of aldosterone. Effective treatment requires supplementation with both Mg and K *(6,25)*. Reduced dietary intake of K is a rare cause of hypokalemia in healthy persons. However, decreased intake may intensify hypokalemia caused by the use of diuretics, for example. GI loss of potassium occurs most often through vomiting, diarrhea, laxative use, gastric suction, or discharge from an intestinal fistula. On average, serum potassium decreases by 0.3 mmol/L for each 100 mmol loss from body stores.

Increased cellular uptake. Alkalemia, insulin, hypothermia, and a variety of drugs (epinephrine, bronchodilators, caffeine, etc.) can increase cellular uptake of potassium *(22)*. When blood is alkaline (alkalemia), intracellular H^+ moves into the blood to normalize extracellular pH. As this happens, both K and Na ions enter cells to maintain electroneutrality. Plasma K decreases by

~0.4 mmol/L (~0.4 mEq/L) per 0.1 unit rise in pH. Additional factors such as HCO_3^- administration, diuretics, and vomiting can intensify the hypokalemia *(6,27)*.

Insulin, which may be stimulated by the administration of glucose, promotes the entry of K ions into skeletal muscle and liver cells. Because insulin therapy may expose or intensify an underlying hypokalemic state, plasma K should be monitored carefully whenever insulin is administered to susceptible patients *(22,26)*.

Symptoms of hypokalemia

Patients are often asymptomatic, especially with mild hypokalemia. The symptoms of hypokalemia are predominantly related to muscular or cardiac function, such as the following:

- Weakness or fatigue
- Muscle cramps, pain, or paralysis may occur in severe cases
- Worsening diabetes control or polyuria
- Palpitations and irritability
- Psychological symptoms such as psychosis, delirium, and depression
- Cardiac abnormalities may be exacerbated by hypokalemia

Muscle weakness, fatigue, or paralysis are probably caused by alterations in the polarization of the cell membrane. Muscle weakness and paralysis may also interfere with breathing.

A variety of cardiac arrhythmias can be induced when potassium falls below 2.5 mmol/L (2.5 mEq/L), which may cause sudden cardiac arrest in some patients:

- Premature atrial and ventricular beats
- Sinus bradycardia
- Atrioventricular block
- Ventricular tachycardia and fibrillation

Laboratory evaluation of hypokalemia

Once hypokalemia is noted, measurement of urinary K concentration can determine if kidney loss of K is a cause. A urine K concentration >15 mmol/L or a urine K:creatinine ratio greater than 13 mmol K/mmol creatinine (115 mmol K/g creatinine) indicate excess urinary loss of K. Next, acid—base status should be assessed because metabolic alkalosis or acidosis will affect plasma K concentrations. Serum Mg should be measured to rule out Mg deficiency *(28)*.

Treatment of hypokalemia

Treatment for hypokalemia should aim at reducing the cause of potassium losses and replacing the lost potassium usually by oral or i.v. replacement.

Especially in severe hypokalemia, other electrolytes may be lost, such as Na^+, Cl^-, Mg^{++}, HCO_3^-, and H^+. An approximate guideline for replacement is for every 1 mmol/L decrease in serum potassium, the potassium deficit is approximately 200–400 mmol. Patients with a plasma potassium of 2.5–3.5 mmol/L may need only oral potassium including eating foods with a high potassium content such as bananas or orange juice. Severe hypokalemia requires aggressive iv replacement. If plasma potassium concentration is less than 2.5 mmol/L, i.v. potassium should be given, with monitoring of ECG and potassium concentrations in blood (27). Potassium chloride or potassium phosphate are the usual salts administered. Potassium bicarbonate may be considered only when metabolic acidosis is present (22). Magnesium concentrations must also be checked because serum potassium is difficult to replenish if the serum magnesium level is also low.

Hyperkalemia

Hyperkalemia is defined as a serum or plasma potassium level greater than about 5.0 mmol/L. If potassium concentrations are only slightly elevated to ~5.5 mmol/L, symptoms are usually not apparent or only mild. Potassium concentrations above 6.5–7.0 mmol/L can lead to cardiac arrythmias or significantly affect hemodynamic and neurologic functions. At concentration above 8.5 mmol/L, life-threatening respiratory paralysis or cardiac arrest can occur and be fatal (29). As with most electrolyte disorders, the rate of change can be more important than the absolute concentration. Patients with chronic hyperkalemia may be asymptomatic at increased levels, while patients with large acute increases in potassium concentrations may develop severe symptoms at lower concentrations (30).

Pseudohyperkalemia is commonly caused by preanalytical effects due to improper patient preparation, specimen collection, or handling, especially if they cause in vitro hemolysis. See Chapter 10 for more details.

Causes of hyperkalemia

The more common causes of true hyperkalemia are shown below and listed in Table 8.6:

- Decreased renal excretion due to acute or chronic kidney disease, adrenal deficiency, diuretics, or other drugs, such as those for treating hypertension.
- Processes that enhance the intracellular release of K^+ into the ECF, such as metabolic acidosis, trauma, rhabdomyolysis, diabetes, low renin or aldosterone, or drugs that block cellular uptake of K^+, such as digoxin (29). Digoxin inhibits the Na/K-ATPase pump that normally causes Na ions to leave cells and K ions to enter cells.
- Excessive oral or i.v. intake of K, rarely causes hyperkalemia in persons with good kidney function.

TABLE 8.6 Causes of hyperkalemia *(31).*

Cellular redistribution	
Metabolic acidosis	Initially causes cellular loss of K. If prolonged, hypokalemia can result.
Insulin deficiency	Increases intracellular movement of K ions
β-Blockers for hypertension (propranolol, atenolol)	Impair cellular uptake of K
ACE inhibitors (captopril, lisinopril, etc)	Suppress angiotensin II leading to decreased aldosterone
Cell injury	Trauma, rhabdomyolysis, hemolytic processes
Decreased kidney excretion	
Oliguric kidney failure	Decreased delivery of Na ions to distal tubule = less K loss in urine
Aldosterone deficiency	Diminishes renal loss of K
K-sparing diuretics	Spironolactone
Defect of cortical collecting tubule	Less K ion is secreted into luminal fluid of nephron.
Excess intake	
Excess dietary intake of K	Almost always requires impaired kidney function that diminishes excretion of K
Pseudohyperkalemia	
Preanalytical effect of in vitro loss of K from erythrocytes, leukocytes, etc.	Cell injury or clotting during blood collection

Patients with hyperkalemia often have an underlying disorder such as renal insufficiency, diabetes mellitus, or metabolic acidosis that, in combination with other processes, causes hyperkalemia *(4,16,29).*

Decreased renal excretion. When glomerular filtration or tubular function is impaired, hyperkalemia is much more likely. For example, during the administration of KCl, a person with decreased kidney function is much more likely to develop hyperkalemia than is a person with normal kidney function.

Because aldosterone is critical for the process of sodium reabsorption and potassium secretion in the collecting duct, diminished aldosterone production caused by Addison's disease or drugs can impair renal excretion of K^+ and cause hyperkalemia *(16,30).*

Several drugs commonly cause hyperkalemia by inhibiting the production or effect of aldosterone, especially in patients with renal insufficiency *(29)*. These drugs include:

- Captopril and other inhibitors of angiotensin-converting enzyme
- Nonsteroidal antiinflammatory agents (inhibit aldosterone)
- Spironolactone (K-sparing diuretic that blocks distal tubular secretion of K)
- Digoxin (inhibits Na-K pump)
- Cyclosporine (inhibits renal response to aldosterone)
- Heparin therapy (inhibits aldosterone synthesis)

Cellular losses or shifts. Potassium may be released into the ECF when tissue breakdown or catabolism is enhanced, resulting in hyperkalemia, especially if renal insufficiency is present *(22)*. Factors that can shift K^+ into extracellular fluid (blood) include the following:

- Trauma and burns
- Metabolic acidosis
- Administration of cytotoxic agents
- Severe tissue hypoxia
- in vivo hemolytic processes and rhabdomyolysis
- Insulin deficiency
- Blood transfusions
- Excessive exercise
- Hyperosmolality

Increased potassium intake. Increased oral intake of potassium rarely causes hyperkalemia in persons with normal kidney function. In healthy persons, an acute oral load of potassium will increase plasma K only transiently, because most of the absorbed K^+ will rapidly move intracellularly. Normal cellular processes gradually release this excess K^+ into the plasma, where it is removed by renal excretion. However, if any degree of kidney insufficiency is present, foods or dietary supplements high in potassium should be avoided. If potassium is infused intravenously at rates in excess of 20 mmol/h, hyperkalemia is possible and especially likely with impaired kidney function *(30)*.

Many of the processes above produce extreme biochemical or physical stresses to cells, which release large amounts of potassium. These stresses include severe trauma, cytotoxic drugs, severe tissue hypoxia, and some hemolytic processes, including blood transfusions. Patients on cardiopulmonary bypass may develop mild elevations in plasma potassium during warming after surgery. Hypothermia increases the movement of potassium into cells, whereas warming causes the cellular release of potassium.

Patients with diabetes are at high risk for hyperkalemia because they typically have multiple causative factors, such as a diet high in potassium and low in sodium (salt substitute), and some degree of kidney disease that may include suppressed renin levels that cause low aldosterone. Also, either insulin deficiency or resistance to insulin limits the ability for cellular uptake of K^+. Hyperglycemia also contributes by producing a hyperosmolar plasma that pulls water and some associated K ions from cells (29).

In metabolic acidosis, H^+ inhibits Na/K-ATPase causing cellular loss of K^+ into plasma. Plasma K increases by 0.2–1.7 mmol/L (0.2–1.7 mEq/L) for each 0.1 unit reduction in pH (29). Because cellular K^+ often becomes depleted in cases of acidosis with hyperkalemia (including diabetic ketoacidosis), treatment with agents such as insulin and bicarbonate can cause a rapid intracellular movement of K^+, causing a rebound hypokalemia.

In blood stored for transfusions, K^+ is gradually released from erythrocytes during storage, which can markedly elevate plasma K concentrations after storage for several weeks. While transfusions have little effect in adults with adequate renal function, infants are more susceptible to hyperkalemia from blood transfusions.

Exercise is uniformly associated with a rise in plasma K^+ concentration, because K^+ is released from cells during exercise. The rise is directly related to the intensity of exercise (4): mild to moderate exercise may increase plasma K^+ by 0.3–1.2 mmol/L (0.3–1.2 mEq/L), while exhaustive exercise may increase plasma potassium by >2.0 mmol/L (>2.0 mEq/L). These changes usually reverse after sufficient rest. Note that forearm exercise during venipuncture can cause erroneously high plasma K^+ concentrations.

Hyperosmolality causes diffusion of water out of cells that carries intracellular K^+ with water into the blood.

Clinical effects of hyperkalemia

The clinical effects of hyperkalemia depend largely on the rate of increase in serum potassium concentrations. Hyperkalemia can cause muscle weakness by decreasing the ratio of intra- to extracellular potassium, which alters neuromuscular conduction. Muscle weakness does not usually develop until plasma potassium reaches 7 mmol/L (7 mEq/L), although the correlation between symptoms and potassium levels varies among patients (22,29,30).

Hyperkalemia disturbs cardiac conduction, which can cause cardiac arrhythmias and possible cardiac arrest. Plasma potassium concentrations of 6–7 mmol/L (6–7 mEq/L) may alter the electrocardiogram, and concentration above 8.5 mmol/L can cause life-threatening respiratory paralysis or cardiac arrest (22,29).

Laboratory evaluation of hyperkalemia

If hyperkalemia is detected or suspected, the ECG should be measured because cardiac abnormalities can be lethal. Simon et al. note that increased plasma K concentrations change ECG in a dose-dependent manner (30):

5.5–6.5 mmol/L = tall t-waves
6.5–7.5 mmol/L = loss of p-waves

7—8 mmol/L = wide QRS complex
8—10 mmol/L = cardiac arrhythmias, sine wave patterns, and asystole

Additional laboratory tests should include serum creatinine and urea and urinalysis to evaluate any kidney disease; urinary Na and K to determine if urinary losses are a cause; serum calcium because hypocalcemia can intensify the cardiac effects of hyperkalemia; blood glucose and blood gas analysis if diabetes or acid—base abnormalities are involved; and LDH and urine myoglobin for possible hemolysis or rhabdomyolysis. If necessary, serum cortisol and aldosterone should be measured to rule out mineralocorticoid deficiency.

Treatment of hyperkalemia

Hyperkalemia should always be confirmed before aggressive treatment in cases where the serum potassium is elevated without explanation. The urgency of therapy depends on the symptoms, the magnitude and rate of increase in serum potassium concentrations, and the cause of the hyperkalemia *(31)*.

For acute hyperkalemia, intravenous calcium solutions rapidly normalize the decreased excitability threshold of myocardial cells caused by the increased potassium concentration. Therefore, intravenous calcium provides immediate but short-lived protection to the myocardium against the effects of hyperkalemia and does not alter the plasma K concentration *(31)*.

Insulin lower plasma potassium by increasing Na-K-ATPase pump activity to promote the entry of K^+ back into cells. Glucose is also administered to prevent hypoglycemia. B-2 receptor agonists (such as albuterol) for asthma also promote the release of insulin. In patients with metabolic acidosis who remain hyperkalemic after the above therapeutic efforts, infusion with sodium bicarbonate may be given to lowers plasma potassium levels *(31,32)*. Patients treated with these agents should be monitored to prevent hypokalemia caused by the overly rapid movement of K^+ into cells.

For chronic hyperkalemia, the judicious approach is to determine if the patient has any degree of kidney failure. Next, determine if the patient is on any medications or dietary supplements that might either impair potassium excretion or contain potassium *(31)*. Thiazide diuretics are preferred if the eGFR is above 30 mL/min, while loop diuretics should be used if kidney impairment is more severe *(29,31)*.

Proper collection and handling of samples

See Chapter 10 for preanalytical causes of artifactual hyperkalemia.

Chloride

Physiology and regulation

As the major extracellular anion, Cl^- functions with Na^+, K^+, and other cations in conduction and transport mechanisms between cells and across cell membranes. Cl^- has several important functions that are sometimes passive responses to other ion movements:

- Cl^- maintains osmotic and fluid balance along with Na^+
- Cl^- is transported along with cations and exchanged with bicarbonate to maintain ionic neutrality
- In acid—base balance, Cl^- typically has an inverse relationship with bicarbonate

Regulation of Cl^- is often an indirect response to the regulation of other ions. Aldosterone increases both Cl^- and Na^+ reabsorption in the renal tubules, with either K^+ exchanging or Cl^- following the Na^+ ion to balance the ionic charge movements (8). A NCC in the kidney, also called a thiazide-sensitive Na/Cl cotransporter, reabsorbs Na and Cl ions from the tubular fluid into the cells of the DCT of the nephron. In the ascending limb of the loop, a Cl^- pump actively reabsorbs Cl^- (16). In the GI tract, Cl^- may passively follow as Na^+ is absorbed, or Cl^- may exchange with HCO_3^- as HCO_3^- is secreted into the intestine. ADH deficiency, by promoting the renal loss of H_2O, leads to increases of Na^+ and Cl^- in blood. Elevated PTH levels are associated with increased serum Cl^- and may play an indirect role in Cl^- regulation (33).

Aldosterone is the key hormone that regulates renal sodium-chloride retention by several mechanisms in the kidney: by NCC and other cotransporters in the DCT; by sodium channels in the collecting duct; and by chloride-bicarbonate exchanging proteins called pendrins in the cortical collecting duct (23). These cotransporters are the target of many drugs.

Cl^- is normally lost through sweat, urine, and gastric secretions. Cl^- movements are related to acid—base balance, with Cl^- and HCO_3^- having an inverse relationship. In Cl^- deficiency, whenever Na^+ is reabsorbed, more HCO_3^- is preferentially reabsorbed instead of Cl^-. With primary aldosterone excess, increased tubular reabsorption of Na^+ and Cl^- is accompanied by loss of both K^+ and H^+.

Sweat chloride measurements have long been used to confirm the diagnosis of cystic fibrosis in children. Use of this test has increased as screening requirements for cystic fibrosis in children have been implemented (16). Because of excess loss of Cl^- in sweat, hypochloremia can occur in cystic fibrosis.

Causes of hypochloremia and hyperchloremia

Hypochloremia is an electrolyte disturbance in which the serum chloride concentration is abnormally low. It is frequently associated with another electrolyte abnormality, such as hyponatremia. Chloride measurements are very useful in evaluating acid—base disorders. In chronic respiratory acidosis, metabolic compensation increases renal HCO_3^- reabsorption and urinary loss of Cl^- *(4,8)*. In metabolic alkalosis, hypochloremia also occurs as a result of increased HCO_3^- reabsorption in the kidney with loss of Cl^- and K^+ *(8)*.

Hypochloremia may be caused by any of the following conditions:

- Metabolic alkalosis caused by vomiting or gastric suctioning leading to excessive loss of gastric HCl.
- Renal loss due to prolonged use of diuretics or to tubular dysfunction that causes excess loss of Cl^- in urine. Hypochloremic metabolic alkalosis can result from these excess losses of Cl^-.
- Diuretics. Loop diuretics, such as furosemide, interfere with tubular reabsorption of Na^+ and Cl^-, which can lead to contraction of the extracellular space. Thiazide diuretics enhance K^+ loss directly.
- Increased aldosterone in response to volume depletion *(8)*.
- In chronic respiratory acidosis, as metabolic compensation increases renal HCO_3^- reabsorption, urinary loss of Cl^- is increased *(4,8)*.
- Contraction alkalosis occurs when a large volume of fluid is lost. Aldosterone is secreted to increase Na^+ and HCO_3^- reabsorption, with Cl-lost in the process *(8)*. This can occur with diuretics, diarrhea, cystic fibrosis, and other conditions *(8,34)*.

Hyperchloremia usually indicates a metabolic acidosis. Gains of chloride by GI absorption or intravenous administration can cause hyperchloremia. As discussed in the Chapter 2 "Physiologic Mechanisms and Diagnostic Approach to Acid-Base Disorders," here are a few principles:

- If the Cl^- concentration goes up while the Na^+ concentration remains the same, an anion such as bicarbonate must decrease.
- May occur with loss of nonchloride sodium salts from GI tract or from urine.
- Will develop with the administration of fluids containing chloride salts.
- Gains or losses of potassium can also affect sodium and chloride concentrations, with hyperchloremia associated with hypernatremia.

Chloride measurements are useful when interpreting difficult acid—base disorders, because hypochloremia indicates a primary or compensatory alkalotic process, whereas hyperchloremia often indicates an acidotic process.

Hyperchloremia is diagnosed when the Cl^- concentration in the blood exceeds 108 mmol/L. Hyperchloremia may occur with excessive loss of bicarbonate due to the following:

- GI losses, such as diarrhea
- Renal tubular acidosis
- Aldosterone deficiency
- ADH secretion can cause HCO_3^- secretion and Cl^- reabsorption in some kidney cells *(8)*
- Compensated respiratory alkalosis
- Drugs, such as cortisone and acetazolamide

Mild elevations may also be seen in primary hyperparathyroidism, where the increase in plasma chloride concentration is associated with a metabolic acidosis caused by the renal tubular effects of parathyroid hormone *(33)*.

Treatment involves diagnosing the underlying cause, whether caused by a disease, a medication, or diet *(35)*. If hypochloremia is mild, it may be corrected by consuming more NaCl salt. For more severe hypochloremia, IV fluids may be given.

Reference intervals for chloride are shown in Table 8.1. Reference intervals for Cl in children are slightly wider than in adults: 96–110 mmol/L *(35)*.

Bicarbonate

Physiology and regulation

HCO_3^- is the second most abundant anion in the extracellular fluid. Total CO_2 concentrations normally range from 21 to 30 mmol/L and is composed of HCO_3^-, carbonic acid (H_2CO_3), and dissolved CO_2, with HCO_3^- accounting for >90% of the total CO_2 at physiological pH.

HCO_3^- is the major component of the buffering system in the blood. In erythrocytes, carbonic anhydrase converts CO_2 to HCO_3^-, which buffers against acidosis by combining with excess H^+ to produce CO_2 and H_2O. The loss of excess CO_2 (an acidic gas) by the alveolar ventilation in the lungs gives the HCO_3^--CO_2 system a great capacity to buffer acid production. In the kidneys, most (85%) of the HCO_3^- is reabsorbed by the proximal tubules, with 15% reabsorbed by the distal tubules.

Causes of decreased and increased bicarbonate

HCO_3^- decreases in metabolic acidosis when HCO_3^- combines with excess H^+ to produce CO_2, which is removed by alveolar ventilation. The typical response to metabolic acidosis is compensation by hyperventilation, which lowers pCO_2 in the blood. Some typical causes of metabolic acidosis are hypoxemia, ketoacidosis, and diarrhea. Other causes of decreased plasma bicarbonate include the following:

- An elevated concentration of protein anions such as albumin can lead to a decreased bicarbonate concentration.
- Hypochloremia. Deficiency of Cl^- in blood enhances HCO_3^- reabsorption in the kidney tubule. When cations such as Na^+ and K^+ are reabsorbed, an anion must follow. Because less Cl^- is available, more HCO_3^- is reabsorbed.
- Contraction alkalosis occurs when a large volume of fluid is lost from the body. To increase blood volume, aldosterone secretion increases reabsorption of Na^+ and HCO_3^- *(8,34)*.

- In respiratory acidosis, to compensate for the excess CO_2 levels, the kidneys reabsorb more bicarbonate *(4,36)*.

In metabolic alkalosis, the elevated plasma bicarbonate is compensated by hypoventilation that increases the pCO_2. Typical causes of metabolic alkalosis include the following *(37)*:

- Severe vomiting
- Hypokalemia
- Excessive alkali intake

Reference ranges for total CO_2 in the blood are shown in Table 8.1.

Self-assessment and mastery

Assess the clinical status of the following patients and their laboratory results for possible disorders of water and/or electrolyte balance.

Case 1. A 56-year-old man diagnosed with small cell lung cancer develops progressive lethargy and confusion. The following laboratory data were obtained:

Na	119 mmol/L (119 mEq/L)
Glucose	6.22 mmol/L (112 mg/dL)
K	4.6 mmol/L (4.6 mEq/L)
BUN	3.2 mmol/L (9 mg/dL)
Cl	77 mmol/L (77 mEq/L)
Serum osmolality	251 mmol/L (251 mOsm/kg)
HCO_3^-	26 mmol/L (26 mEq/L)
Urine osmolality	857 mmol/L (857 mOsm/kg)

What is the most likely cause of this patient's abnormal sodium? Is the calculated osmolality consistent with the measured osmolality?

Case 2. A 21-year-old woman with insulin-dependent diabetes was brought to the hospital in a coma. On admission, she was hypotensive (blood pressure 90/20 mmHg) with a rapid pulse (122/min) and rapid breathing (32/min). Her laboratory data were as follows:

Na	134 mmol/L (134 mEq/L)
K	6.4 mmol/L (6.4 mEq/L)
Glucose	66.1 mmol/L (1200 mg/dL)
pH	6.80
pCO_2	10 mmHg
HCO_3^-	3 mmol/L (3 mEq/L)

What is the likely cause of these abnormalities? What interventions would you expect to be done? What would be the effect of this therapy?

Case 3. A 40-year-old woman with edema is found on exam to be hypertensive, with the following laboratory data:

Na	145 mmol/L (145 mEq/L)
K	2.8 mmol/L (2.8 mEq/L)
Cl	106 mmol/L (106 mEq/L)
HCO_3^-	30 mmol/L (30 mEq/L)
Arterial pH	7.48
Urine K	50 mmol/L (50 mEq/L) (abnormally high for hypokalemia)

What does the combination of elevated sodium with decreased potassium suggest in this patient?

Case 4. A 62-year-old man with congestive heart failure and mild renal failure was treated with the angiotensin-converting enzyme inhibitor captopril to relieve the congestive heart failure. The following lab results were obtained:

Na	134 mmol/L (134 mEq/L)
K	6.8 mmol/L (6.8 mEq/L)
Cl	102 mmol/L 102 mEq/L)
HCO_3^-	21 mmol/L (21 mEq/L)
BUN	10.0 mmol/L (28 mg/dL)
Creatinine	159.12 μmol/L (1.8 mg/dL)
Glucose	6.38 mmol/L (115 mg/dL)

After further treatment with sodium polystyrene sulfonate, plasma K decreased to 4.8 mmol/L (4.8 mEq/L).

What caused the initial elevated potassium and the later normalized potassium?

Case 5. An elderly man was evaluated for having nausea and vomiting for 4 weeks. He was taking no drugs and had no signs of hypovolemia, with a normal blood pressure and pulse, and his skin turgor was normal appearing. His renal function was also normal. His lab results on plasma were as follows:

Na:	125 mmol/L
K:	4.4 mmol/L
Cl:	97 mmol/L
Total CO_2:	20 mmol/L
Glucose:	75 mg/dL
Creatinine:	0.9 mg/dL
pH:	7.44

Aldosterone and cortisol were both very low.

Evaluate this patient for possible explanations of his low Na, normal K, and low aldosterone.

Case 6. A **serum** K level is ordered in a patient with thrombocytosis and a platelet count of 10×10^5/mm^3 (normal 1.5−4.5 x 10^5/mm^3). The serum K level is 6.3 mmol/L. Considering both clinical and preanalytical possibilities, what would be the next logical action for the physician to take?

a. Give the patient calcium gluconate
b. Administer glucose and insulin
c. Draw another blood sample for a serum K level
d. Draw a blood sample for a plasma K level
e. Put the patient on renal dialysis

Self-assessment and mastery discussion

Assessment of Case 1. This patient had the classic symptoms of the syndrome of inappropriate secretion of ADH (SIADH): decreased serum sodium, decreased serum osmolality, and increased urine osmolality. The lower-normal BUN is also typical of SIADH.

Small cell lung cancers often produce a peptide with ADH-like activity. Usually, acute therapy for SIADH is simply to restrict water intake. However, because of the more severe symptoms indicating cerebral edema, the patient was treated with hypertonic saline and furosemide (a diuretic). The diuretic was given to prevent hypervolemia.

The calculated osmolality is 247 mOsm/kg:

$$(2 \times 119) + 3.2 + 6.22 = 247$$

which is close to the measured osmolality of 251 mmol/kg.

Assessment of Case 2. This patient was in severe diabetic ketoacidosis. She was given 8 U of insulin immediately and 8 U/h thereafter. The insulin decreased glucose at ~5.55 mmol/L (100 mg/dL) per hour. Insulin will also lower the plasma potassium. She also received 4 liters of IV normal saline over the first hour. Patients with severe diabetic ketoacidosis are often profoundly volume depleted for a number of reasons, including glucose induced diuresis, poor oral intake, vomiting, and insensible losses from increased minute ventilation. Providing insulin and IV fluids is typically adequate to stabilize a patient such as this. Rarely, sodium bicarbonate infusion may be required to increase the pH and provide much-needed additional buffer capacity because the patient is nearly depleted. Increasing the pH will also lower the potassium concentration. It is imperative to monitor vital signs and electrolytes very closely during the first few hours of treatment for severe diabetic ketoacidosis (case adapted from ref. *13*).

Assessment of Case 3. The plasma value for sodium and potassium suggests mineralocorticoid excess, most likely caused by primary hyperaldosteronism. The combination of hypertension, hypokalemia, metabolic alkalosis, and inappropriately increased urinary potassium (urinary K should be decreased with hypokalemia) suggests a possible diagnosis of primary hyperaldosteronism (case adapted from ref. *3*).

Assessment of Case 4. Captopril is a vasodilator that diminishes angiotensin by inhibiting angiotensin-converting enzyme. Although captopril is effective in the treatment of congestive heart failure, hyperkalemia may result possibly from the inhibition of aldosterone secretion. Sodium polystyrene sulfonate is an ion-exchange resin that lowers plasma K by binding to K in the gut. In this case, it brought the plasma K down to 4.8 mmol/L (4.8 mEq/L).

Assessment of Case 5. This patient has two disorders that could affect the K and other electrolytes in opposite directions. The low aldosterone should cause a low Na, an elevated K and Cl, and possibly a mild acidosis. His long history of vomiting should cause an alkalosis, with an elevated total CO_2, a decreased K, and a decreased Cl. The best explanation is that this person has multiple disorders that have offset each other and made the interpretation more difficult. These two conditions have combined to produce hyponatremia, with a normal K, and nearly normal Cl, bicarbonate, and pH. The very low cortisol and aldosterone suggest Addison's disease. As mentioned in the text, while an elevated K is associated with Addison's disease, about one-third of cases have a normal plasma K.

Assessment of Case 6. While actions a, b, and e will lower a blood K level, both a and e are for more drastic hyperkalemia and b may not be necessary. Since a serum K was ordered, the high platelet count suggests that the clotting process may have released some K into the serum and caused the modest hyperkalemia. Thus, another serum K would likely give a similar result, and getting a plasma K should prove that the patient's K level is actually normal, or (if plasma K still elevated) would confirm the hyperkalemia. Answer **d** is the best answer.

References

1. Nebelkopf Elgart, H.; Johnson, K. L.; Munro, N. Assessment of Fluids and Electrolytes. *AACN Clin. Issues* **2004,** *15,* 607−621.
2. Armstrong, L. E. Assessing Hydration Status: The Elusive Gold Standard. *J. Am. Coll. Nutr.* **2007,** *26,* 575S−584S.
3. *Fluid and Electrolyte Balance.* www.mcb.berkely.edu/courses/mcb135e/kidneyfluid.html.
4. Narins, R. G., Ed. *Maxwell and Kleeman's Clinical Disorders of Fluid and Electrolyte Metabolism,* 5th ed.; McGraw-Hill: New York, 1994.
5. Sharma, R.; Sharma, S. Physiology, Blood Volume. In *StatPearls;* NCBI Bookshelf, January 2020. https://pubmed.ncbi.nlm.nih.gov/30252333.

6. Rose, B. D. *Clinical Physiology of Acid-Base and Electrolyte Disorders,* 3rd ed.; McGraw-Hill: New York, 1989.

7. Henderson, R.; Bonsall, A. *Osmolality, Osmolarity and Fluid Homeostasis;* Patient Info, July 2016. Hoorn EJ, Halperin ML, Zietse R. Diagnostic approach to the patient with hyponatremia: traditional versus physiology-based options. Q J Med 2005; 98: 529-540. https://patient.info/doctor/osmolality-osmolarity-and-fluid-homeostasis.

8. Wagner, C. A. Effect of Mineralocorticoids on Acid-Base Balance. *Nephron. Physiol.* **2014,** *128,* 26−34.

9. Grider, M. H.; Jessu, R.; Glaubensklee, C. S. *Physiology, Action Potential;* NCBI Bookshelf, January 2020. https://www.ncbi.nlm.nih.gov/books/NBK538143.

10. Lewis, J. L., III. *Overview of Sodium's Role in the Body.* Consumer Version; Merck Manual, April 2020. https://www.merckmanuals.com/home/hormonal-and-metabolic-disorders/electrolyte-balance/overview-of-sodiums-role-in-the-body.

11. Mistry, A. C.; WynneBM, Y. I.; Tomlin, V.; Yue, Q.; Zhou, Y.; et al. Ths Sodium Chloride Cotransporter (NCC) and Epithelial Sodium Channel (ENaC) Associate. *Biochem. J.* **2016,** *473* (19), 3237−3252.

12. Simon, E. E.; Batuman, V. *Hyponatremia;* Medscape eMedicine, June 2019. https://emedicine.medscape.com/article/242166-overview.

13. Waikar, S. S.; Mount, D. B.; Curhan, G. C. Mortality after Hospitalization with Mild, Moderate, and Severe Hyponatremia. *Am. J. Med.* **September 2009,** *122* (9), 857−865.

14. Lukitsch, I.; Batuman, V. *Hypernatremia;* Medscape eMedicine, January 2021. https://emedicine.medscape.com/article/24109.

15. Tsipotis, E.; Price, L. L.; Jaber, B. L.; Madias, N. E. Hospital-Associated Hypernatremia Spectrum and Clinical Outcomes in an Unselected Cohort. *Am. J. Med.* **2018,** *131* (1), 72−82.e1.

16. Scott, M. G.; LeGrys, V. A.; Hood, J. L. Electrolytes and Blood Gases. In *Tietz Textbook of Clinical Chemistry and Molecular Diagnostics;* Burtis, C. A., Ashwood, E. R., Bruns, D. E., Eds., 5th ed.; Elsevier Saunders: St Louis, 2012; pp 807−835.

17. Spasovski, G.; Vanholder, R.; Allolio, B.; Annane, D.; Ball, S.; Bichet, D.; et al. Clinical Practice Guideline on Diagnosis and Treatment of Hyponatraemia. *Nephrol. Dial. Transplant.* **2014,** *29* (Suppl. 2), i1−i39.

18. Knochel, J. P. Hypoxia is the Cause of Brain Damage in Hyponatremia. *J. Am. Med. Assoc.* **1999,** *281,* 2342−2343.

19. Decaux, G.; Musch, W. Clinical Laboratory Evaluation of the Syndrome of Inappropriate Secretion of Antidiuretic Hormone. *Clin. J. Am. Soc. Nephrol.* **2008,** *3,* 1175−1184.

20. Kapoor, D.; Srivastava, M.; Singh, P. Point of Care Blood Gases with Electrolytes and Lactates in Adult Emergencies. *Int. J. Crit. Illn. Inj. Sci.* **2014,** *4* (3), 216−222.

21. Shah, K.; Khardori, R. *Hyponatremia in Emergency Medicine;* Medscape eMedicine, August 2020. https://emedicine.medscape.com/article/767624.

22. Weiss-Guillet, E.-M.; Takala, J.; Jakob, S. M. Diagnosis and Management of Electrolyte Emergencies. *Best Pract. Res. Clin. Endocrinol. Metab.* **2003,** *17,* 623−651.

23. Weiner, I. D.; Wingo, C. S. Chapter 38: Endocrine Causes of Hypertension—Aldosterone. In *Comprehensive Clinical Nephrology,* 4th ed.;; 2010; pp 469−476.

24. Sedlacek, M.; Schoolwerth, A. C.; Remillard, B. D. Electrolyte Disturbances in the Intensive Care Unit. *Semin. Dial.* **2006,** *19,* 496−501.

25. Palmer, B. F.; Clegg, D. J. Physiology and Pathophysiology of Potassium Homeostasis: Core curricUlum 2019. *Am. J. Kidney Dis.* **2019,** *74* (5), 682−695.

26. Castro, D.; Sharma, S. Hypokalemia. In *StatPearls;* NCBI Bookshelf, July 2020. https://www.statpearls.com/articlelibrary/viewarticle/23269.

27. Lederer, E.; Batuman, V. *Hypokalemia;* Medscape eMedicine, January 2021. https://emedicine.medscape.com/article/242008.

28. Castro, D.; Sharma, S. Hypokalemia. In *Statpearls;* NCBI Bookshelf, July 2020. https://www.ncbi.nlm.nih.gov/books/NBK482465.

29. Lederer, E.; Batuman, V. *Hyperkalemia;* Medscape eMedicine, April 2020. https://emedicine.medscape.com/article/24090.

30. Simon, L. V.; Hashmi, M. F.; Farrell, M. W. Hyperkalemia. In *Statpearls;* NCBI Bookshelf, December 2020. https://www.ncbi.nlm.nih.gov/books/NBK470284.

31. Palmer, B. F.; Clegg, D. J. Diagnosis and Treatment of Hyperkalemia. *Cleve. Clin. J. Med.* **2017,** *84* (12), 934−942.

32. Garth, D.; Schraga, E. D. *Hyperkalemia in Emergency Medicine Clinical Presentation;* Medscape eMedicine, October 2018. https://emedicine.medscape.com/article/766479.

33. Lafferty, F. W. Primary Hyperparathyroidism: Changing Clinical Spectrum, Prevalence of Hypertension, and Discriminant Analysis of Laboratory Tests. *Ann. Intern. Med.* **1981,** *141,* 1761−1766.

34. Brinkman, J. E.; Sharma, S. Physiology, Metabolic Alkalosis. In *Stat Pearls;* 2020. www.ncbi.nlm.nih.gov/books/NBK482291/.

35. Falck, S.; Seladi-Schulman, J. Hypochloremia: What Is It and How Is It Treated? *Healthline* **September 29, 2018**. https://www.healthline.com/health/hypochloremia.

36. Hamm, L. L.; Nakhoul, N.; Hering-Smith. Acid-Base Homeostasis. *Clin. J. Am. Soc. Nephrol.* **2015,** *10,* 2232−2242.

37. Khanna, A.; Kurtzman, N. A. Metabolic Alkalosis. *J. Nephrol.* **2006,** (Suppl. 9), S86−S96.

Chapter 9

Lactate physiology and diagnostic evaluation

Introduction

By 1900, many physicians had noted metabolic acidosis in patients who were critically ill. In 1925, Clausen identified increased lactate along with acidosis in these patients, which gave rise to the condition of "lactic acidosis," although later work showed the lactate and acid were produced in entirely separate biochemical reactions, as described recently in a review (1). In 1964, Broder and Weil observed that lactate levels above 4 mmol/L were associated with poor outcomes in patients with shock, as cited in a review by Andersen et al. (2). While a decreased supply of oxygen to cells caused by inadequate cardiac, pulmonary, or circulatory function is well recognized as a cause of increased blood lactate, other causes have more recently become appreciated, such as mitochondrial dysfunction. Together, these causes of hyperlactatemia have led to very large increases in lactate testing in critical care.

Over the past 35 years, the measurement of blood lactate has evolved from a test regarded as having little clinical value to a highly valuable tool in monitoring general metabolic function. The annual test volume at Duke Medical Center for blood lactate has increased from approximately 2000 tests in 1985, to 10,000 in 1995, to 30,000 in 2005, and is now over 60,000. Reasons for this increase started in 1986 when a pediatric cardiac surgeon began to use blood lactate for monitoring the status of pediatric patients during open-heart surgery. This continued with use during extracorporeal membrane oxygenation (ECMO) procedures (3), and further with use in evaluating patients in the ED, including trauma patients and those with chest pain, and as a criterion for admission to a higher level of care. Over the last 10–15 years, major increases in lactate testing have resulted from official guidelines recommending lactate measurement for evaluating sepsis and for monitoring the effectiveness of therapy in patients with sepsis (4).

Physiology and metabolism

Lactate is converted from pyruvate in the cytoplasm of most cells, with the highest levels produced in muscle cells. In the process of glycolysis, which is

Blood Gases and Critical Care Testing. https://doi.org/10.1016/B978-0-323-89971-0.00005-7

193

always an anaerobic process, glucose is converted to pyruvate. In cells with no mitochondria such as erythrocytes, glycolysis is the only pathway for producing ATP. With an adequate supply of oxygen to cells with functional mitochondria, pyruvate is converted to acetyl CoA by pyruvate dehydrogenase with the essential cofactor thiamine, then enters the Krebs cycle and is eventually metabolized by the very complex process of oxidative phosphorylation to produce CO_2 and large amounts of ATP. Under these conditions, a relatively small amount of pyruvate is converted to lactate, so blood lactate remains within normal intervals. Whether pyruvate is converted to Acetyl CoA or to lactate depends on the proportions of NAD^+ and NADH. With adequate oxygen in the mitochondria, more NAD^+ is produced, which favors the conversion of pyruvate to Acetyl CoA. However, if oxygen supply is inadequate, or in cells with no mitochondria (such as erythrocytes), levels of NADH are increased, which favors the conversion of pyruvate to lactate by the enzyme lactate dehydrogenase. These processes are shown in Fig. 9.1. Contrary to common belief, the production of lactate from pyruvate does not produce acid, but actually consumes acid in the reaction:

$$\text{Pyruvate} + \text{NADH} + H^+ = \text{Lactate} + NAD^+ \qquad (9.1)$$

The continual production of NAD^+ is necessary for the production of ATP, either by the highly efficient process of oxidative phosphorylation in mitochondria or from the inefficient process in the cytosol where pyruvate is converted to lactate.

This anaerobic production of lactate is far less efficient, ultimately producing much less ATP and large amounts of lactate, which diffuses into the

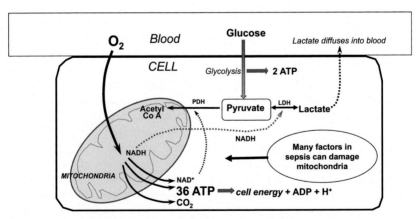

FIGURE 9.1 Relationships of glucose, pyruvate, oxygen, NAD^+, and NADH in the production of lactate. This shows that the production of lactate from pyruvate depends directly on the ratio of NAD^+/NADH, and indirectly depends on the supply of oxygen. Adequate oxygen must be supplied and metabolized by mitochondria to ensure continued conversion of NADH to NAD^+. Without sufficient oxygen and functional mitochondria, NADH accumulates and favors production of NAD^+ to NADH by converting pyruvate to lactate.

blood. Hydrogen ions also accumulate from the degradation of ATP to ADP in the reaction:

$$ATP = ADP + PO_4 + H^+ \qquad (9.2)$$

Thus, the concentration of lactate in the blood becomes a clinical marker for numerous pathologic processes such as cardiac and circulatory insufficiency (shock, trauma, coagulopathy, etc.), sepsis, pulmonary disease or dysfunction, and various types of drug or toxin overdoses, including exposure to carbon monoxide or cyanide.

Lactate is mainly metabolized by the liver (around 60%) and kidney (around 30%), and the rate of lactate clearance in the normal liver exceeds the rate of lactate production of other tissues *(5)*. Because it is usually associated with damage to the liver, kidney, or mitochondria, diminished clearance of lactate often indicates a poorer outcome *(2)*.

Causes of hyperlactatemia

Lactate concentrations are commonly elevated in acutely ill patients, notably in sepsis and septic shock, trauma, cardiac and pulmonary insufficiencies, during and after cardiac surgery, and other types of ischemia due to cardiac arrest or circulatory problems *(2)*. Consequently, blood lactate measurements are frequently used to evaluate the initial condition of the patient and to monitor the effectiveness of therapy. In addition to the traditional view that elevated blood lactate is caused by an oxygen deficit to tissues, elevated lactate can also result from mitochondrial dysfunction. While this includes the rare cases of cyanide poisoning, mitochondrial dysfunction is an important contributor in sepsis, where several factors associated with sepsis inhibit mitochondrial function and elevate blood lactate: inflammation, cytokines, platelet and endothelial activators, tissue necrosis factor, etc. Some common uses of blood lactate measurements include the following:

- In neonates during and after open-heart surgery for congenital heart disease *(3)*;
- To evaluate patients who may require ECMO and for monitoring their progress;
- For triage in an ED setting to determine which patients need more immediate care and for monitoring the effectiveness of therapy;
- In trauma patients, early identification of increased blood lactate followed by aggressive resuscitation improves survival. Although published nearly 30 years ago, the principles of lactate interpretation have stood the test of time. Survival was very high (98%–100%) in patients whose blood lactate normalized within 24 h; was 75%–80% in patients whose blood lactate normalized in 24–48 h; but relatively poor in patients whose blood lactate could not be normalized by 48 h *(6)*. A study of intubated trauma patients in the ED found that both serum lactate and end-tidal pCO$_2$ correlated with hospital mortality *(7)*.

- During open-heart surgery, patients are cooled to reduce oxygen consumption. However, this can also cause perfusion abnormalities (vasoconstriction and shunting) that can lead to tissue hypoxia. Along with the anesthesia and drugs, these can cause problems with oxygen metabolism that leads to increased blood lactate. Thus, blood lactate measurements have become a means to monitor such patients.
- Pulmonary embolism, often associated with deep-vein thrombosis, is a major cause of hospitalization or mortality, with elevated blood lactate levels correlated to both a high mortality rate and prothrombotic fibrin properties (8).
- Identification of high-risk ICU patients needing more aggressive therapy. Patients who responded to treatment by reduction in lactate to less than 1.0 mmol/L, had a mortality less than 7%.
- In cirrhotic patients, blood lactate typically increases with the severity of cirrhosis. The increased blood lactate appears related to accelerated glycolysis in the splanchnic region coupled with reduced capacity to metabolize lactate in the liver (9).
- In goal-directed therapy for sepsis, blood lactate measurements (lactate >2.0 mmol/L) have become an essential component of "sepsis bundles" for detecting higher-risk patients (10,11). Blood lactate is elevated in some patients with sepsis who have no evidence of hypoperfusion or tissue hypoxia. As noted earlier, several factors associated with sepsis can elevate lactate, including inflammation, cytokines, platelet and endothelial activators, and tissue necrosis factor.
- As a summary statement, blood lactate concentrations evaluate the complex metabolic state of the patient experiencing surgery, trauma, sepsis, anesthesia, hypothermia, some drugs, inflammation, coagulopathies, etc.

Table 9.1 lists several causes of elevated blood lactate concentrations (2).

Blood lactate in sepsis

As noted earlier, several factors associated with sepsis inhibit mitochondrial function and elevate blood lactate: inflammation, cytokines, platelet and endothelial activators, tissue necrosis factor, etc. Blood lactate can be used as a marker for evaluating and monitoring the complex circulatory, cellular, and metabolic disturbances that occur in sepsis patients. Monitoring blood lactate is now included in the "sepsis bundles" used as guidelines for improving outcomes in sepsis and septic shock. These bundles have evolved over the years, with the most important revision in 2018 that the 3-h and 6-h bundles have been combined into a single "Hour-1 bundle" (4,11). Sepsis is now defined as life-threatening organ dysfunction caused by a dysregulated host response to infection. Organ dysfunction in sepsis may now be evaluated by an increase in the Sequential Organ Failure Assessment (SOFA) score of two points or more (4).

Septic shock is defined as a subset of sepsis in which particularly profound circulatory, cellular, and metabolic abnormalities are associated with a greater risk of mortality than with sepsis alone. Patients with septic shock can be

TABLE 9.1 Some clinical causes and examples of elevated blood lactate concentrations.

Cause	Examples
Shock	Hypovolemic, sepsis-related
Cardiac insufficiency	Myocardial infarction, congestive heart failure, cardiac arrest
Respiratory failure	Pulmonary edema, obstructive lung disease, severe hypoxemia
Tissue ischemia	Trauma, burns, gut, other organs
Drugs or toxins	Alcohol, cocaine, carbon monoxide, cyanide.
Pharmacologic drugs	Metformin, propofol, acetaminophen, linezolid, theophylline.
Hyperactivity of muscles	Seizures, excessive work of breathing, intense exercise.
Mitochondrial diseases	Diseases that uncouple oxidative phosphorylation and cause destruction and leakage of mitochondrial DNA and peptides. May be related to sepsis, myopathies, toxins, and other causes.
Liver failure	Cirrhosis, acute liver diseases that cause delayed clearance of lactate.
Thiamine deficiency	Thiamine is a cofactor for PDH, the enzyme that converts pyruvate to acetyl-CoA.

clinically identified by a vasopressor requirement to maintain a mean arterial pressure (MAP) $\geq$65 mmHg and lactate >2 mmol/L (>18 mg/dL) in the absence of hypovolemia (4). Table 9.2 summarizes the changing definitions of sepsis, and Fig. 9.2 shows the progression of a systemic infection to sepsis and septic shock along with the associated changes in lactate and other physiologic parameters.

Clinical approach to monitoring blood lactate

For a variety of clinical situations, the guidelines for measuring blood lactate are similar. Whether on trauma patients, following surgery, in suspected sepsis, or with unexpected changes in a patient's condition, lactate should be measured immediately, then evaluated to determine if continued and more frequent monitoring is needed, as follows:

Measure lactate as soon as possible:

Lactate normal: GOOD
Lactate slightly elevated: initiate therapy
Lactate markedly elevated: consider more aggressive therapy

TABLE 9.2 Summary of the changing definitions of sepsis.

Sepsis category	SEP-1/SEP-2 (~2005)	SEP-3 (2016)
Sepsis	2 of 4 SIRS criteria AND suspected infection	SOFA score >2 + suspected infection
Severe sepsis	Sepsis + organ failure + hypoperfusion or hypotension	*This category has been eliminated*
Septic shock	Lactate >4 mmol/L Systolic BP <90 mmg or MAP <70 mmHg and not responsive to fluids	Vasopressors required to maintain MAP above 65 mmHg, and lactate >2 mmol/L without hypovolemia

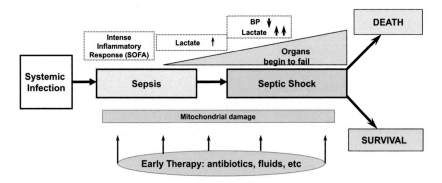

SOFA = Sequential Organ Failure Assessment

FIGURE 9.2 Progression of infection to sepsis and septic shock, based on SEP-3 guidelines. Sepsis begins with a systemic infection that progresses to an intense inflammatory response that may lead to septic shock. As these processes develop, mitochondrial dysfunction occurs and organs begin to fail, with a possible progression to death. Blood lactate begins to rise in these processes and may continue to increase along with a drop in blood pressure if sepsis progresses to septic shock.

Measure lactate every 3—24 h:

Lactate decreasing: GOOD
Lactate staying elevated: increase level of therapy
Lactate rising: BAD — consider more aggressive therapy

Evaluate after 24 h:

Lactate normal or close to normal: GOOD
Lactate still clearly elevated: consider most aggressive therapy

Basic treatment strategies for elevated blood lactate

Because multiple conditions can contribute to elevated blood lactate, one must carefully evaluate the patient's medical history, physical assessment, and results of laboratory or diagnostic tests. As with blood gas and electrolyte disorders, treatment of elevated blood lactate levels should be aimed at correcting the underlying cause. If the cause is related to hypoperfusion, hypoxemia, or hypotension (shock state), the goal should be to improve perfusion and blood pressure by administering fluids, vasopressors, or inotropes as appropriate. If caused by drugs, seizures, malignancy, or thiamine deficiency, the goals should be to stop, reverse, and treat the offending condition *(2)*.

Sepsis has several guidelines aimed at rapid and early treatment that emphasize prompt recognition and early management with fluid resuscitation and antibiotic administration. The "sepsis bundles" have evolved over the years to an hour-1 bundle that emphasizes the importance of immediate treatment, especially in patients with hypotension *(11)*.

Within 1 h bundle:

- Measure lactate, then remeasure if >2 mmol/L.
- Obtain blood cultures.
- Administer broad-spectrum antibiotic, then refine as culture results become available.
- Begin rapid administration of fluids (30 mL/kg) for hypotension or lactate >4 mmol/L.
- Consider vasopressor administration if the patient remains hypotensive during or after fluid resuscitation, with the goal to maintain MAP >65 mmHg.

Depending on the circumstances, other actions may be the following:

- Order blood counts, urinalysis, CT scans, X-rays, procalcitonin, etc. as appropriate.
- Reassess volume status and tissue perfusion.
- Repeat lactate if warranted
- Consider invasive mechanical ventilation if acute respiratory distress syndrome is present.

Proper collection and handling of specimens for lactate testing

See Chapter 10.

Reference intervals for blood lactate

Reference intervals for blood lactate differ among publications *(2)* and are slightly different between arterial and venous blood. Furthermore, interpretation has become more complicated with sepsis guidelines now recommending that a blood lactate >2.0 mmol/L requires follow-up testing, as this level can be a marker for poorer outcomes *(12)*. The following are general guidelines for reference intervals and interpretations:

Arterial blood lactate: <1.8 mmol/L for normal adult at rest
Venous blood lactate: <2.2 mmol/L for normal adult at rest
Blood lactate from 2.2 to 4.0 mmol/L is moderately elevated

Blood lactate above 4.0 mmol/L is seriously elevated, although the clinical situations must be evaluated because the direction of change is important. In the first example, for a blood lactate of 4 mmol/L or above, if the elevation is because the patient has just undergone surgery, this lactate is not necessarily serious. If the lactate is increasing in trauma, surgery, or in patients with metabolic disorders, despite full medical intervention (cardiac, pulmonary, fluid, antibiotic, etc., support), a lactate above 4 mmol/L represents a more serious situation. In the second example, if a patient's lactate is 4.2 mmol/L but is decreasing following therapy from a high of 6.0 mmol/L (for example), that would be of lesser concern, especially if the lactate concentration continues to decrease.

Self-assessment

Self-assessment questions

1. Which of the following do not favor optimal production of ATP?
   a. A high ratio of NADH to NAD+
   b. Well-functioning mitochondria
   c. An adequate supply of oxygen to mitochondria
   d. Continual production of NAD+
   e. Conversion of pyruvate to acetyl-CoA
2. Which of the following is a feature of the SEP-3 guidelines:
   a. Meeting the SIRS criteria, plus an infection
   b. SOFA score >2, plus an infection
   c. Severe sepsis
   d. Blood lactate >4 mmol/L
   e. Mean arterial pressure >80 mmHg
3. Which is likely the least serious cause of an elevated lactate?
   a. Pulmonary embolism
   b. Cirrhosis of the liver
   c. Intense exercise
   d. Exposure to carbon monoxide
   e. Sepsis

4. Which of the following time sequence of blood lactate results (mmol/L) is likely the most serious?

Choice	Lactate Time 0	Lactate 6 hours	Lactate 24 hours	Lactate 60 hours
a	4.2	3.5	2.0	1.1
b	1.8	2.0	2.2	1.5
c	1.8	4.2	3.0	2.0
d	1.8	2.8	2.8	2.8

5. Which is not an appropriate intervention in management of a patient with sepsis?
a. Order blood lactate right away
b. Administer broad-spectrum antibiotics right away
c. Remeasure lactate if initial lactate is 3.0 mmol/L
d. Immediately put the patient on a ventilator
e. Administer fluids if MAP is <40 mmHg

Answer Key:

1. a
2. b
3. c
4. d
5. d

Case examples

Case 1: A 24-year-old patient was one of six arriving in the ED following a multivehicle accident. At triage, the patient was in moderate pain, so lab tests, X-rays, and CT scans were ordered. Blood gases and electrolytes were normal, and the venous lactate was 1.4 mmol/L. The X-ray and CT scans revealed a minor fracture in the forearm. The patient was given medication for pain and the fracture was eventually set, while other more seriously injured patients received urgent care. The normal lactate gave the physician assurance that the patient was likely not seriously injured, and allowed other more seriously injured patients to be cared for.

Case 2. Following coronary artery bypass surgery, a 56-year-old male had an aortic balloon pump inserted through the right femoral artery to maintain sufficient cardiac output. Two hours postsurgery, blood lactate was 3.2 mmol/L. Several parameters were rechecked to determine the possible cause. Cardiac output was good, there was no evidence of gut ischemia, and the airways were clear. However, poor pulses were noted in the right leg. They felt this could be due to the balloon pump constricting blood flow to the leg, so the balloon pump was removed. Two hours later the lactate was normalizing at 1.5 mmol/L. This illustrates the usefulness of lactate measured on a patient whose surgery was uncomplicated and had no obvious problems. The elevated lactate was a warning that something was not right and warranted further investigation.

Case 3. A 23-year-old male was admitted to the ED with a gun-shot wound to his side. Because the wound was small, bleeding was minimal, and he had stable vital signs, he was not believed to need emergent surgery. However, lab tests came back with arterial pH = 7.23, pCO_2 = 40 mmHg, HCO_3 = 18 mmol/L, and lactate = 3.8 mmol/L. Because of these results, the patient was reevaluated and a CT scan was ordered. The CT scan showed the bullet had apparently hit the patient's stomach, small intestine, and spleen. He was immediately taken to surgery for repair of these injuries and was discharged after 3 days. This is an example where external signs suggested minimal damage, while blood lactate and other laboratory tests indicated more serious internal damage, which was confirmed by the CT scan.

References

1. Gunnerson, K. J.; Pinsky, M. R. *Lactic Acidosis,* September 2020. https://emedicine.medscape.com/article/167027.
2. Andersen, L. W.; Mackenhauer, J.; Roberts, J. C.; Berg, K. M.; Cocchi, M. N.; Donnino, M. W. Etiology and Therapeutic Approach to Elevated Lactate Levels. *Mayo Clin. Proc.* **2013,** *88* (10), 1127–1140.
3. Toffaletti, J.; Hansell, D. Interpretation of Blood Lactate Measurements in Paediatric Open-Heart Surgery and in Extracorporeal Membrane Oxygenation. *Scand. J. Clin. Lab. Invest.* **1995,** *55,* 301–307.
4. Singer, M.; Deutschman, C. S.; Seymour, C. W.; et al. The Third International Consensus Definitions for Sepsis and Septic Shock (Sepsis-3). *J. Am. Med. Assoc.* **2016,** *315* (8), 801–810.
5. Cheng, C.-Y.; Kung, C.-T.; Wu, K.-H.; et al. Liver Cirrhosis Affects Serum Lactate Level Measurement While Assessing Disease Severity in Patients with Sepsis. *Eur. J. Gastroenterol. Hepatol.* **2020;** https://doi.org/10.21203/rs.3.rs-16937/v1.
6. Abramson, D.; Scalea, T. M.; Hitchcock, R.; Trooskin, S. Z.; Henry, S. M.; Greenspan, J. Lactate Clearance and Survival Following Injury. *J. Trauma* **1993,** *35* (4), 584–589.
7. Safari, E.; Torabi, M. Relationship between End-Tidal CO2 (ETCO2) and Lactate and Their Role in Predicting Hospital Mortality in Critically Ill Trauma Patients; A Cohort Study. *Bull. Emerg. Trauma* **2020,** *8* (2), 83–88.
8. Zabczyk, M.; Natorska, J.; Janion-Sadowska, A.; Malinowski, K. P.; Janion, M.; Undas, A. Elevated Lactate Levels in Acute Pulmonary Embolism are Associated with Prothrombotic Fibrin Clot Properties: Contribution of NETs Formation. *J. Clin. Med.* **2020,** *9,* 953.
9. Jeppesen, J. B.; Mortensen, C.; Bendtsen, F.; Moller, S. Lactate Metabolism in Chronic Liver Disease. *Scand. J. Clin. Lab. Invest.* **2013,** *73* (4), 293–299.
10. Levy, M. M.; Dellinger, R. P.; Townsend, S. R.; Linde-Zwerble, W. T.; Marshall, J. C.; Bion, J.; et al. The Surviving Sepsis Campaign: Results of an Internationals Guideline-Based Performance Improvement Program Targeting Severe Sepsis. *Intensive Care Med.* **2010,** *36,* 222–231.

11. Levy, M. M.; Evans, L. E.; Rhode, A. The Surviving Sepsis Campaign Bundle: 2018 Update. *Intensive Care Med.* **2018**, *44,* 925—928.

12. Wacharasint, P.; Nakada, T.-A.; Boyd, J. H.; Russell, J. A.; Walley, K. R. Normal-Range Blood Lactate Concentration in Septic Shock is Prognostic and Predictive. *Shock* **2012,** *38* (1), 4—10.

Chapter 10

Collection and handling of samples: effects on blood gases, Na, K, ionized Ca, Mg, lactate, and phosphate analyses

Sources of preanalytical errors in blood gas and electrolyte testing

Introduction

There are numerous sources of preanalytical errors in laboratory testing. Among these are the following:

- Wrong test ordered
- Wrong patient collected
- Error in sample collection, including improper patient preparation
- Errors during specimen transport
- Errors during specimen processing

This chapter will address only the last three of these. As noted by Karon *(1)* and Azman et al. *(2)*, the top five reasons for redrawing specimens are shown in Table 10.1.

Clearly, hemolysis and cell leakage or rupture are major causes of preanalytical errors. Contamination of a sample could include improper anticoagulant, air bubbles in the sample and contamination with IV fluids, such as saline, anticoagulants, or drugs.

Hemolysis

In vitro hemolysis is by far the most common preanalytical problem in clinical laboratory testing. It causes problems in (a) preventing hemolysis during collection, (b) detecting hemolysis in blood samples, and (c) both interpreting

Blood Gases and Critical Care Testing. https://doi.org/10.1016/B978-0-323-89971-0.00011-2

TABLE 10.1 Most frequent reasons for redrawing blood from patients in the emergency department (ED).

Reason for redraw from ED	% of ED redraws
Hemolysis	70
Specimen clotted	11
Some contamination of specimen suspected	4
Sample never received in lab	2

results and deciding whether to report the results for many tests, most notably potassium.

Preventing in vitro *hemolysis.* To prevent hemolysis, several guidelines may be followed:

1. Use a 23 gauge or larger needle. Smaller gauge needles can damage RBCs passing through the needle.
2. Collecting blood through an IV infusion line often causes a higher rate of hemolysis.
3. When mixing blood with anticoagulant, gently invert the tube. Never shake the tube.
4. For several reasons, blood collection tubes should be filled to their stated capacity.
5. Never forcefully expel blood from a syringe into a tube, especially an evacuated tube.

Detecting hemolysis: plasma/serum or whole blood. In former years, hemolyzed serum or plasma in evacuated tubes was usually detected by the watchful eye of skilled clinical laboratory scientists who manually handled centrifuged blood specimens. They would then grade the level of hemolysis, such as slight, moderate, or gross, then follow the laboratory's protocol for whether to report none, some, or all of the results, or request a new specimen be collected. However, visual detection of hemolysis is subject to individual interpretation and variability. Furthermore, elevated bilirubin, a common occurrence in neonates, can further complicate the visual interpretation of hemolysis *(2)*.

As automation has become common in central laboratories, the samples are centrifuged, analyzed, and often reported without visual inspection of the plasma or serum. This issue has largely been solved by modern analyzers that use spectrophotometry to detect hemolysis and assign a numerical grade to its severity to alert the laboratory if the level of hemolysis significantly affects any test results. A highly debated issue is whether to report a result that is urgently requested by the physician *(2)*. This is also a significant concern for any testing done at point-of-care where whole blood is analyzed and hemolysis would not likely be detected by the user. There is one commercial system that advertises a separate device that is able to detect hemolyzed blood at the point of care (www.hemcheck.com).

Detecting hemolysis is a significant concern for blood gas samples that are collected in syringes and must be analyzed immediately and without centrifugation. Thus, visual detection of hemolysis is not a practical option before analyzing the sample. A less than ideal option is to rapidly centrifuge the specimen after the analysis if either the results (i.e., an elevated K) or the patient's previous samples suggest hemolysis is present.

As yet, there are no blood gas analyzers that have built-in systems to detect hemolysis in uncentrifuged blood during analysis, although companies are likely working on systems that detect hemolysis during the testing process. Much of the technology is still proprietary, but the basic technology used for detection is likely based on either: (1) isolating plasma from the flow system during or before the measurement process and then using optical methods to detect levels of hemolysis; or (2) using an algorithm based on multiple analyte inputs to predict the degree of hemolysis in whole blood. While the first is similar to that used in automated clinical chemistry analyzers, the novel approach for blood gas analyzers would focus on in-line, real-time plasma isolation to determine a sample's hemolysis index. The second may have limited reliability in its reliance on whole blood alone, given the known pathological variations and nonlinear relationship between analyte changes in the presence of in vitro hemolysis. As blood gas testing continues to shift to the point of care, whatever real-time detection system of hemolysis is used, it must be both reliable and simple and have flexibility in how to alert clinicians to the impact on critical results such as potassium.

In vivo hemolysis occurs when erythrocytes are ruptured in circulation. Causes include immune reactions with cells, hemolytic anemias, and mechanical rupture from cardiac bypass, ECMO, or heart valve devices. While in vitro hemolysis that falsely affects test results is much more common and accounts for 98%−99% of hemolyzed specimens, results on samples with in vivo hemolysis will likely give a true physiologic increase of analytes in the blood. Thus, laboratory results such as potassium are appropriate and should be reported. Detecting in vivo hemolysis presents other challenges as detection often requires either a patient's history of a hemolytic process or one or more prior hemolyzed samples from the patient.

Reporting results on hemolyzed samples. Usually, the laboratory does not report any results that are significantly affected by in vitro hemolysis and requests that another specimen be collected, preferably by another person collecting the blood. The more difficult issue is when a suitable specimen cannot be recollected and the physician calls the laboratory with an urgent need for the result, which may provide some assurance in difficult clinical situations. Legal wisdom says to not report the results under any circumstances, while accommodating medical urgency says to report the result as a comment with disclaimers such as noting the result is significantly affected by hemolysis and is provided only at the request of the physician and after discussion with the laboratory director. An approach that addresses both concerns is to require one or two additional redraws, then if both specimens are still unacceptable, the laboratory provides the results if the physician requests them.

Proper collection and handling of samples

Blood gases

Blood collected for blood gas analysis is susceptible to changes, especially to pO_2. Anaerobic conditions during collection and handling are essential because room air has a pCO_2 of nearly 0 and a pO_2 of $\sim$150 mmHg. The factors that must be controlled are:

- Removal of all air bubbles
- Use of the proper anticoagulant
- Appropriate use of plastic syringes (glass syringes rarely used anymore)
- The temperature of storage before analysis
- The length of delay between collection and analysis of blood
- Any agitation of the specimen

The complete removal of all air bubbles is especially important before sending blood in a syringe by pneumatic transport, which will agitate the sample and markedly affect pO_2, with pCO_2 much less affected by air bubbles *(3,4)*.

The effect of liquid heparin at <10% (vol./vol.) of the volume of blood has variable effects on blood gas and other analytes *(5,6)*. There appears to be little effect on pH and a relatively small effect ($\sim$3%) on pO_2, although the pO_2 also appears to be affected by the pO_2 in the heparin diluent *(5,6)*. pCO_2, bicarbonate, and base excess are affected proportionately by dilution, about a 10% decrease. As expected, liquid heparin will dilute other constituents in blood, such as electrolytes, lactate, and glucose, which are often analyzed simultaneously in many current blood gas and electrolyte analyzers *(7)*. Therefore, only dry heparin should be used as an anticoagulant.

Although plastic syringes are used for nearly all blood gas measurements, they have a potential disadvantage because they are permeable to oxygen and absorb oxygen in polyethylene *(8)*. When stored in ice, because of the increased affinity of Hb for O_2 at cold temperatures, blood can absorb oxygen within the wall of the syringe that has diffused through the plastic. This effect is most pronounced in samples with a pO_2 of $\sim$100 mmHg and above; that is, when Hb is already nearly fully saturated with oxygen and is unable to buffer any added

O_2. A pO_2 of 100 mmHg may increase by 8 mmHg during 30 min of storage on ice. When Hb is less saturated (e.g., at a pO_2 of 60 mmHg), it is better able to buffer the additional oxygen, causing a relatively small change in pO_2.

Collection and transport of blood for oxygen measurements

As the pO_2 in the blood is easily changed, there are several important pre-analytical cautions necessary to prevent errors in clinical measurements of oxygen and cooximetry results.

Delay in sample processing. In general, storage of blood in plastic syringes at room temperature is acceptable if the analysis is within 15 min, with average changes in pH of <0.01 unit, pCO_2 of less than 1 mmHg, and pO_2 of less than 2 mmHg *(9)*. For room temperature storage up to 60 min, the rates of change for pH and pCO_2 are small. However, for pO_2, storage at ambient temperature for >30 min decreased pO_2 by an average of about 5 mmHg, with a wide variation of changes *(9,10)*. In a study of samples with original pO_2 values of 50 and 250 mmHg, storage for 15 min at ambient temperature (22−24°C; 72−75°F) decreases pO_2 by an average of ∼5 mmHg after 15 min, and by ∼8 mmHg after 30 min *(11)*.

Samples from patients with extremely high leukocyte and/or platelet counts must be analyzed as soon as possible because pH and pCO_2 can change to some degree, while pO_2 and glucose and lactate can change dramatically in samples stored at room temperature *(12)*. Even storage on ice did not prevent significant changes in pO_2 in such samples. For such samples, point-of-care testing using either blood gas analyzers or pulse oximeters offer blood gas and oximetry results that are much less affected by cell metabolism.

Storage time of specimens after collection: Room temperature versus icing. Although storage of blood at room temperature can lower pO_2, pH, and glucose, and increase pCO_2 and lactate, prolonged storage of plastic syringes in ice can also increase pO_2. Because polyethylene is slowly permeable to oxygen (polypropylene is the membrane used in pO_2 electrodes) and because cold significantly increases the affinity of Hb for oxygen, prolonged storage of blood in plastic syringes can cause a slow diffusion of O_2 into the blood within the syringe. More of this oxygen binds to the Hb at the cold temperatures, then is released when the sample is analyzed at 37°C *(8)*. As an example, a pO_2 of 100 mmHg may increase by 8 mmHg if stored on ice for 30 min. As a general guide, if a specimen will be analyzed in 30 min or less, do not ice the specimen. If analysis will be delayed longer than 30 min, the syringe should be placed in ice. Because of increased leakage from cells, storage on ice can also increase the potassium concentration in the stored blood *(13)*.

Insufficient line waste draw. Arterial and venous catheters must be adequately flushed before drawing the blood specimen for blood gas testing. As a general rule, the waste draw volume should be 2-times the catheter dead volume *(14)*.

Inadequate mixing of heparinized blood. There are at least two important reasons to properly mix blood specimens after they are collected in a syringe: to dissolve and mix the heparin anticoagulant, and to maintain an even distribution of the blood cells.

- There are several different types of dry preparations of heparin used in syringes that may have distinctly different mixing requirements. If heparin is not distributed and dissolved rapidly, clots and/or microclots may form in the specimen.
- Erythrocytes sediment rapidly in undisturbed specimens. If erythrocytes are not evenly distributed by mixing, significant errors can occur in the hemoglobin measurements. As a rule, specimens that have sat for over 5 min may require 2 min of mixing *(15)*. Note that sedimentation is even higher in inflammatory conditions, such as infection or autoimmune disease, as this is the basis of the erythrocyte sedimentation rate (Sed Rate) test.

Dilution by liquid heparin. While the use of liquid heparin in syringes has been almost totally replaced by dry preparations of heparin, if any liquid heparin remains in a syringe before blood collection, it will dilute out the analytes to be tested, especially hemoglobin, electrolytes, glucose, lactate, and pCO_2.

Trapped air in syringes. Trapped air bubbles in syringes can significantly increase or lower the pO_2, %O_2Hb and sO_2, and O_2 content of the blood. Because atmospheric air has a pO_2 of approximately 150 mmHg, exposure to air will increase the pO_2 of blood from patients breathing room air. However, for patients on O_2-enriched air, the pO_2 of blood can rapidly decrease upon exposure. The variables on this effect are as follows:

a. The volume of air trapped in the syringe.
b. The number and size of the air bubbles.
c. Any agitation of the specimen after collection, especially from pneumatic transport.
d. The original pO_2 of the specimen.
e. The original Hb concentration and %O_2Hb of the specimen.

Pneumatic transport of specimens. As noted earlier, because pneumatic transport of blood gas specimens can agitate blood specimens, any air bubbles trapped in the syringe will be equilibrated with the blood and cause significant changes in the pO_2 result *(3,4)*. It is especially important for those collecting blood gas specimens to remove ***all*** air bubbles from the syringe before sending by pneumatic transport. As noted, the pO_2 result will be increased in samples from patients breathing atmospheric air at pO_2 ~ 150 mmHg. However, for blood specimens from patients on supplemental oxygen with a pO_2 such as 300 mmHg, the pO_2 result will be decreased *(4)*.

Collection of capillary samples. Although properly arterialized capillary blood can yield pH and pCO_2 results that are close to arterial results, the agreement of pO_2 results between capillary and arterial blood is more variable *(16,17)*. pO_2 results from capillary blood will always be lower than arterial blood, possibly only slightly lower, so that a normal or low-normal capillary pO_2 assures that the patient is not hypoxemic. The puncture site (finger or heel) should be prewarmed up to 42°C to increase blood flow several fold. After puncture, the blood should be allowed to flow freely, without "milking," which can introduce venous blood and interstitial fluid. The capillary tube should be filled completely without air bubbles and properly sealed *(18)*.

CLSI has recently published a document on collecting capillary blood specimens *(19)*. It describes optimal procedures in the collection process for capillary blood, such as how to approach, greet, and identify the patient, and how to select the proper collection site including which sites to avoid, such as sites of infection or inflammation, fingers of newborns, and patients with either edema or severe dehydration. Proper positioning of the patient is described with collection techniques, along with labeling and transporting the specimen. Analytical variation between capillary and venous specimens is discussed, such as noting that total protein, calcium, bilirubin, sodium, and chloride are lower by 5% or more in capillary blood compared to venous blood. There are several topics covered related to quality assurance, such as safety, equipment management, handling unexpected events, and continual improvement *(19)*.

Collection of umbilical cord (UC) blood specimens. There are two general techniques for collecting UC blood specimens. In Method 1, the cord blood is collected while the UC is still attached, and within 1 min after delivery of the neonate and before the placenta is released. This has the advantage of maximizing UC blood returning to the neonate.

In Method 2, a segment of the UC is clamped off within 1 min after delivery of the neonate. After clamping, blood may be collected from the clamped segment of the UC.

Blood from the UC artery should be collected first, then blood from the UC vein. The UC artery has tougher walls and is a bit more challenging to collect *(20)*. Because of this, the needle should be inserted into the UC artery at an angle (see Fig. 10.1) to avoid accidently collecting venous blood instead of arterial blood. Collecting from the correct UC artery or vein is usually not a problem unless the blood volume in the UC is low *(21)*.

There are a few results that may indicate an improper collection of UC blood:

1. A small difference in pH and pCO_2 between the UC arterial and venous blood. While this is not an uncommon occurrence with valid results on UC arterial and venous blood due to impairment of placental exchange, it often occurs when both UC "arterial and venous" blood samples are drawn from the same vessel, usually the UC vein.

FIGURE 10.1 Proper technique for collecting UC arterial and venous blood. Note the importance of inserting the needle at an angle to avoid puncturing through the intended blood vessel. *Original drawing by Madison Willett, a medical technologist in the Clinical Pediatric Laboratory at Duke Medical Center. Used with permission.*

2. An unexpected difference between the UC artery and venous blood. As examples, the UC arterial blood should always have a lower pH and pO_2, and a higher pCO_2 than the UC venous blood.
3. An UC arterial pO_2 greater than about 38 mmHg. According to one study on mothers receiving supplemental oxygen, no UC arterial pCO_2 was >38 mmHg *(22)*.

Sodium

Heparin. While some older preparations of heparin at concentrations >150 U/ mL have been reported to interfere with Na measurements, no effect has been noted for more recent syringes and evacuated tubes that contain <50 U/mL. A negative interference is still possible with underfilled collection syringes or tubes that increase the concentration of heparin (23).

Direct versus indirect measurements of sodium. Several reports have noted differences between sodium results by direct ISE methods compared to methods that dilute the sample in the analysis ("indirect" methods) (24).

More recently, the effects of decreased plasma protein concentrations have been noted to falsely increase sodium results in diluted ISE methods. While the opposite effect is also noted for increased protein concentrations, elevated plasma protein concentrations are much less common in ICU patients, as noted elsewhere: 73% decreased, 26% normal, and less than 1% increased (24,25).

The effects of hemolysis on sodium, potassium, and calcium results are shown in Table 10.2 (26).

Potassium

Not only is potassium one of the more frequently ordered tests in the clinical laboratory, it arguably has the most issues with spurious results due to pre-analytical variables (27). Most of these are related to cellular leakage or rupture because the intracellular concentration of K is about 25 times higher than the extracellular (blood) concentration.

- Patient preparation: the patient should be as calm as possible and avoid excessive fist clenching. The tourniquet should be released by 1 min after application (27). Prolonged tourniquet application can also cause hemo-concentration that spuriously increases other analytes. Muscle contraction by excessive fist clinching can increase K by 0.4−0.8 mmol/L and even higher in specific patients (28).
- Lysis of red cells, WBCs, or platelets: As noted earlier, cell lysis is the most frequent cause of erroneous elevations in K concentrations (1,2).

TABLE 10.2 Changes in test results caused by degree of hemolysis.

Level of hemolysis	Effect on test results (mmol/L)		
	Potassium	Sodium	Calcium
None detected	4.0	140	1.20
0.5% 0.08 g/dL Hb	4.5	140	1.20
5% 0.8 g/dL Hb	7.0	136	1.10
10% 1.5 g/dL Hb	10.0	133	1.00

Hemolysis and cell lysis may be caused by traumatic venipuncture, using a too small needle for phlebotomy, collection via an indwelling catheter, forceful collection into a syringe, or forceful transfer of blood from a syringe to an evacuated tube.

- Clotting: In the clotting process, platelets release a relatively small amount of K into the serum. This usually increases the serum K concentration by about 0.1–0.3 mmol/L. However, with high platelet counts, the effect can be higher.
- High leukocyte levels: If WBC counts are in the range of 100,000 or higher, they can have a dramatic effect on K concentrations in plasma. This effect appears to be less in both serum and in heparinized blood analyzed on blood gas analyzers. In such samples with very high WBCs, I have observed normal potassiums of ∼4 mmol/L, as measured on undiluted heparinized blood on a blood gas analyzer, have a result of 6–10 mmol/L on heparinized plasma analyzed as a diluted sample on a chemistry analyzer. This may be caused by the detergents in diluted chemistry methods lysing WBCs that remain in the plasma supernate, even after centrifugation.
- Time and temperature of storage: If a whole blood specimen is stored at room temperature, K can increase by 0.2 mmol/L after 1.5 h. If stored at 4°C, the increase is much greater (∼0.5 mmol/L per hour) because the chilling of blood inhibits the ATPase ion pumps that will increase leakage of K ions out of erythrocyte and other cells *(29)*.

Ionized calcium

Effects of anticoagulants and clotting on ionized calcium concentrations

Accurate determination of ionized calcium concentration in blood is affected by several preanalytical causes. It is well known that common anticoagulants such as EDTA, citrate, and oxalate function as anticoagulants by binding to calcium ions that will significantly decrease the ionized calcium concentration. While heparin is the common anticoagulant for many chemistry tests, ordinary heparin chelates calcium ions in proportion to the concentration of heparin in the collection device *(30)*. For many years, electrolyte-balanced or other special preparations of dry heparin have been available that minimize this interference, as shown in Fig. 10.2 *(31)*. While some have assumed that serum, having no anticoagulant, has no effect on ionized Ca, Fig. 10.2 shows that the clotting process itself has variable effects on the ionized calcium concentration *(31)*.

Because pH affects the binding of Ca ions to albumin and other proteins, changes in pH will affect the ionized calcium concentration in blood. If the sample has been collected in a heparinized evacuated tube, it must be kept

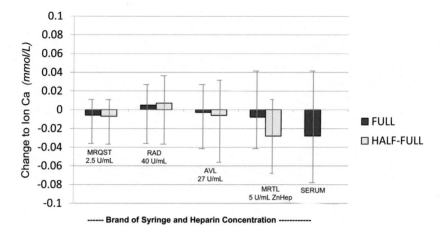

FIGURE 10.2   Changes to ionized calcium concentrations from blood with various anticoagulants and from serum. Changes are relative to whole blood containing no anticoagulant analyzed immediately after collection. Serum was prepared from blood collected in an evacuated tube with no anticoagulant after clotting and centrifugation. *This is an original drawing.*

sealed until analysis to avoid any pH change caused by the loss of CO_2. It is also important to analyze the samples as soon as possible, preferably within 30 min, to minimize the effects of acid production by erythrocyte and leukocyte metabolism *(32)*. Table 10.3 lists the approximate acceptable limits for storage of blood samples for ionized calcium measurements in various types of specimen containers. This is a table used at Duke Clinical Laboratories as a useful, if not perfect, guide for evaluating the acceptability of a variety of specimens received for ionized calcium measurement.

Physiological and preanalytical changes to ionized calcium concentrations

When collecting samples for either total or ionized calcium measurements, tourniquet application should be brief (<1 min) before a specimen is collected. The combined effects of hemoconcentration (causing hyperproteinemia) and localized lactate and acid production can alter total calcium by as much as 10% *(33)* and ionized calcium by ~2%−3% *(34)*. Because loss of CO_2 will increase pH, all samples for ionized calcium measurements must be collected anaerobically. The metabolic activity of cells during storage affects both pH and ionized calcium. If analyzing heparinized plasma, blood should be centrifuged within 1−2 h to prevent acidosis from affecting the ionized calcium concentration. For each 0.1 unit decrease in pH, ionized Ca changes by approximately 0.036 mmol/L *(35)*.

Because of dilutional effects, no liquid heparin anticoagulants should be used. Most heparin anticoagulants (sodium, lithium) partially bind to calcium

TABLE 10.3 Acceptable limits for ionized calcium stability in various specimen containers.

Specimen container	Time elapsed between collection time and analysis time				
	<30 min	30–60 min	1–2 h	2–8 h	>8 h
Unopened syringe	OK	OK	OK[a]	OK[b]	NO
Syringe briefly opened	OK	OK	NO	NO	NO
Unopened vacutainer > $\frac{1}{2}$ full	OK	OK	OK[a]	OK[b]	NO
Unopened vacutainer < $\frac{1}{2}$ full	OK	OK	Report pH corrected ion Ca	Report pH corrected ion Ca	NO
Opened vacutainer > $\frac{1}{2}$ full	OK 10 min	NO	NO	NO	NO
Opened vacutainer < $\frac{1}{2}$ full	OK 2 min	NO	NO	NO	NO
Heparinized capillary tubes	OK	NO	NO	NO	NO

[a]Specimen not on ice; Result may be compromised.
[b]Specimen [2-8] hours old; Result may be compromised.

and lower ionized calcium concentrations. However, syringes containing dry heparin products are available that essentially eliminate the interference by heparin (31,36). These include heparin that has been "balanced" with electrolytes that minimize the binding of calcium to heparin in the blood sample and minimal heparin (~3 U/L) that is dispersed in a soluble inert carbohydrate "web." Furthermore, incomplete filling of syringes containing these heparins has minor effects on the ionized calcium and magnesium concentration (31).

Other causes of preanalytical changes to ionized calcium concentrations include:

- Physical activity: A 5%–10% increase in ionized calcium concentrations was observed in healthy volunteers after bicycling exercise (37). This may be related to increased acid production and decreased bicarbonate.
- Patient posture at the time of sampling could affect ionized calcium concentrations by changes in albumin and protein levels.

- Circadian rhythm: Variations of up to 10% in ionized calcium concentrations have been noted during a 24 h-cycle, with the nadir occurring in the late afternoon *(38)*.
- Ventilatory rate and pH: Because pH affects ionized calcium concentrations in blood by its impact on albumin/protein binding of calcium ions, a rise in pH (lower H ion concentration) lowers ionized calcium concentration by increasing binding of calcium ions by albumin. Conversely, a drop in pH (higher H ion concentration) will increase ionized calcium concentrations. Thus, hyperventilation-induced respiratory alkalosis before sampling may decrease the ionized calcium concentration *(39)*.
- Albumin concentrations will affect ionized calcium concentrations, with increased albumin causing decreased ionized calcium and vice versa *(40)*.

For analysis of total calcium in urine, an accurately timed urine collection is preferred. The urine may be acidified to prevent precipitation of calcium salts during storage of the urine.

Lactate

Lactate in the blood is among the analytes more susceptible to preanalytical factors that can increase the lactate concentration *(26)*. These factors are described in the following:

Patient preparation. Before collecting blood for lactate analysis, the patient should be resting and seated for at least 5 min and avoid any prior strenuous exercise for at least 30 min. Any vigorous hand squeezing to make venipuncture easier should also be avoided.

Temperature of storage. At room temperature storage, whole blood increases lactate concentrations by 0.3–0.5 mmol/L after 30 min, with storage on ice slowing this rate to about 0.1 mmol/L after 60 min. Lactate in plasma is quite stable, with no detectable change after 120 min on ice and changing by 0.1 mmol/L after 120 min at room temperature *(41,42)*.

Use of anticoagulants with antiglycolytic agents versus use of heparin. Although fluoride and/or oxalate (Ox) may interfere with some electrochemical methods for lactate, these preservatives (especially fluoride) minimize increases in lactate to very low rates. In whole blood samples containing F/Ox at room temperature, the mean increase was 0.15 mmol/L over 24 h. Even in samples with high white cell counts, lactate increased by only 0.3 mmol/L over 8 h. This suggests that samples collected in F/Ox will be stable for at least 8 h at room temperature *(42)*.

Because fluoride interferes with some lactate-sensing electrodes, heparinized syringes are nearly always used for critical care testing. Because heparin provides no inhibition to lactate production in either venous or arterial whole blood, lactate should be measured within 15 min after collection. While storage on ice minimizes lactate generation in vitro, ice storage affects other analytes, such as increasing potassium and pO_2.

Duration of storage. Several studies have shown that lactate in heparinized blood increases by an average of about 0.01 mmol/L per minute at room temperature, which shows that such samples should be analyzed within 15–20 min *(41)*. Storage of heparinized blood on ice was more stable, changing by 0.1 mmol/L per hour. Storage of separated plasma was reasonably stable, changing by less than 0.1 mmol/L after 1 h at RT, and by 0.1 mmol/L after 2 h on ice *(41)*.

As Fig. 10.3 shows, there is some variability among samples, which can be related to cell counts.

High cell counts, especially leukocytes. In a study of 110 routinely collected blood samples, lactate concentrations showed a significant positive relationship between blood cells (mainly leukocytes) and the rate of lactate increase at room temperature. The influence of blood cells on lactate production showed the rate of change in concentration was also related to time and temperature: The change in lactate was significantly higher at room temperature (0.012 mmol/L/min) than at 4°C (0.0035 mmol/L/min) ($P < .0001$). A delay in the processing of a whole blood sample of more than 15 min at room temperature or an hour at 4°C overestimates the lactate concentration by more than 0.2 mmol/L. Leukocytosis with cell counts higher than 6×10^{10} cells/L decreased the stability period to only 10 min *(43)*.

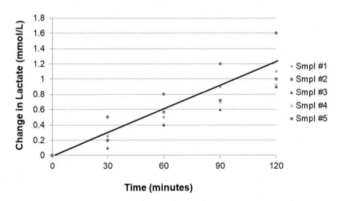

FIGURE 10.3 Lactate changes in five samples of heparinized blood kept at room temperature. This shows the average change is about 0.3 mmol/L per 30 min, but changes do vary from about 0.2 to 0.4 mmol/L per 30 min. *This is an original drawing.*

Phosphate

Because the intracellular concentration of free phosphate (~ 10 mmol/L), is several fold higher than in blood ($\sim 0.8-1.5$ mmol/L), any type of cell lysis or cell leakage will increase the concentration of PO_4 in plasma or serum. As such, hemolyzed specimens should not be used for clinical measurements. Phosphate concentrations in plasma are reported to be about 0.1 mmol/L lower than in serum, and anticoagulants such as citrate, oxalate, and EDTA interfere with spectrophotometric methods based on phosphomolybdate (44). Because prolonged storage of unspun blood at RT or 37°C will increase phosphate concentrations in serum or plasma, a general practice is to separate serum or plasma from cells within 2 h after collection. Hemolyzed specimens are not suitable for storage because the organic phosphate esters contained in erythrocytes can be hydrolyzed during storage. Once separated from cells and stored at 4°C, PO_4 concentrations are stable in serum or plasma for several days (44).

Magnesium

If possible, the patient should fast before the collection of blood. As with phosphate, the intracellular concentration of magnesium is about three-fold higher than in blood ($\sim 0.65-1.05$ mmol/L), so any type of cell lysis or cell leakage will increase the Mg concentration in plasma or serum. As such, serum or plasma should be separated or isolated (if using a serum separator gel) from the clot as soon as possible, and samples with more than mild hemolysis are generally not acceptable.

Anticoagulants such as EDTA, citrate, or oxalate will bind Mg ions and lower the Mg concentration, especially so for ionized Mg methods. Depending on the method, icterus or lipemia may also interfere.

Preanalytical variables for ionized magnesium are similar to those for ionized calcium, which are affected by sample pH and anticoagulants or other additives in the blood collection tube. pH affects ionized Mg inversely: for each 0.1 pH unit change, the ionized Mg changes by about 0.012 mmol/L (35). Ordinary sodium heparin decreases ionized magnesium by chelation, with ionized magnesium decreasing by about 0.01 mmol/L for each 25 IU/mL of heparin (45). Balanced heparins do not have this effect. Citrate lowers the ionized magnesium concentration markedly (46), and silicone found in some, but not all, blood collection devices apparently increases the ionized magnesium result (45).

Self-assessment questions

1. Which test is the most challenging for detecting *in vitro* hemolysis?
   a. Lactate dehydrogenase
   b. Lactate
   c. Blood gases
   d. Calcium
   e. Magnesium

2. Which blood gas test is most sensitive to air bubbles in the specimen?
   a. pCO_2
   b. pH
   c. CO-Hb
   d. O_2-Hb
   e. pO_2
3. True-False: Icing is always a good practice for stabilizing specimens for blood gas analysis.
4. True-False: A pO2 result of 86 mmHg on a properly-collected capillary blood sample can be used for clinical interpretation.
5. True-False: If the pH results on umbilical arterial and umbilical venous blood are the same, it confirms that the blood samples were collected improperly.
6. True-False: Lysis or leakage from white blood cells and/or platelets can alter potassium results as much as lysis of erythrocytes (red blood cells).
7. True-False: Because all anticoagulants have a significant effect on ionized calcium results, serum is the best choice for an accurate measurement.
8. True-False: Using liquid heparin as an anticoagulant for blood gas testing will have the greatest effect on the pH result.
9. Which sample of plasma/serum from the following blood specimens will likely give the most accurate lactate result?
   a. Blood with no anticoagulant left at room temperature for 45 minutes
   b. Blood with no anticoagulant stored on ice for 120 min
   c. Blood collected with heparin and stored for 30 min at room temperature.
   d. Blood collected in a tube containing fluoride/oxalate left at room temperature for 60 min.

Answer Key:

1. c
2. e
3. False
4. True
5. False
6. True
7. False
8. False
9. d

References

1. Karon, B. *Preanalytical Variables,* August 26, 2019.
2. Azman, W. N. W.; Omar, J.; Koon, T. S.; Ismail, T. S. T. Hemolyzed Specimens: Major Challenge for Identifying and Rejecting Specimens in Clinical Laboratories. *Oman Med. J.* **2019,** *34* (2), 94–98.
3. Astles, J. R.; Lubarsky, D.; Loun, B.; Sedor, F. A.; Toffaletti, J. G. Pneumatic Transport Exacerbates Interference of Room Air Contamination of Blood Gas Samples. *Arch. Pathol. Lab Med.* **1996,** *120,* 642–647.
4. Toffaletti, J. G.; McDonnell, E. H. *Effect of Small Air Bubbles on Changes in Blood pO$_2$ and Blood Gas parameters: Calculated vs Measured Effects,* July 2012. http://acutecaretesting. org.
5. Hutchinson, A. S.; Ralston, S. H.; Dryburgh, F. J.; Small, M.; Fogelman, I. Too Much Heparin: Possible Source of Error in Blood Gas Analysis. *Br. Med. J.* **1983,** *287,* 1131–1132.
6. Ordog, G. J.; Wasserberger, J.; Balasubramaniam, S. Effect of Heparin in Arterial Blood Gases. *Ann. Emerg. Med.* **1985,** *14* (3), 233–238.
7. Chhapola, V.; Kuman, S.; Goyal, P. Is Liquid Heparin Comparable to Dry Heparin for Blood Gas Sampling in Intensive Care Unit? *Indian J. Crit. Care Med.* **2014,** *18* (1), 14–20.
8. Mahoney, T. J.; Harvey, J. A.; Wong, R. J.; Van Kessel, A. L. Changes in Oxygen Measurements When Whole Blood is Stored in Iced Plastic or Glass Syringes. *Clin. Chem.* **1991,** *37,* 1244–1248.
9. Rajasekaran, R.; Arthur, H. S.; Peter, J. V. Arterial Blood Gas Tensions – Effect of Storage Time and Temperature. *Indian J. Resp. Care Med.* **2014,** *3* (1), 350–356.
10. Knowles, T. P.; Mullin, R. A.; Hunter, J. A.; Douce, F. H. Effects of Syringe Material, Sample Storage Time, and Temperature on Blood Gases and Oxygen Saturation in Arterialized Human Blood Samples. *Respir. Care* **2006,** *51* (7), 732–736.
11. Beaulieu, M.; Lapointe, Y.; Vinet, B. Stability of pO$_2$, pCO$_2$, and pH in Fresh Blood Samples Stored in Plastic Syringe with Low Heparin in Relation to Various Blood-Gas and Hematological Parameters. *Clin. Biochem.* **1999,** *32* (3), 101–107.
12. Prasad, K. N.; Manjunath, P.; Priya, L.; Sasikumar, S. Overcoming the Problem of Pseudohypoxemia in Myeloproliferative Disorders: Another Trick in the Bag. *Indian J. Crit. Care Med.* **2012,** *16* (4), 210–212.
13. Narayanan, S. The Preanalytical Phase – An Important Component of Laboratory Medicine. *Am. J. Clin. Pathol.* **2000,** *113,* 429–452.
14. Rickard, C.; Couchman, B.; Schmidt, S. A Discard Volume of Twice the Deadspace Ensures Clinically Accurate Arterial Blood Gases and Electrolytes and Prevents Unnecessary Blood Loss. *Crit. Care Med.* **2003,** *31* (6), 1654–1658.
15. Narayanan, S. *Preanalytical Issues Related to Blood Sample Mixing,* October 2005. www. acutecaretesting.org.
16. Cousineau, J. *Neonate Capillary Blood Gas Reference Values,* January 2006. www. acutecaretesting.org.
17. McLain, B. I.; Evans, J.; Dear, P. R. F. Comparison of Capillary and Arterial Blood Gas Measurement in Neonates. *Arch. Dis. Child.* **1988,** *63,* 743–747.
18. Higgins, C. *Capillary Blood Gases – to Arterialize or Not,* July 2008. www.acutecaretesting. org.
19. CLSI. Collection of Capillary Blood Specimens. In *CLSI Standard GP42,* 7th ed.; Clinical and Laboratory Standards Institute, 2020.
20. Higgins, C. *Umbilical-Cord Blood Gas Analysis,* October 2014. www.acutecaretesting.org.

21. Jorgensen, J. S. *Umbilical-Cord Blood Gas Analysis in Obstetrical Practice;* Radiometer Webinar, July 2015.

22. Saneh, H.; Mendez, M. D.; Srinivasan, V. N. *Cord Blood Gas;* NCBI Bookshelf, April 20, 2020. www.ncbi.nlm.nih.gov/books/NBK545290.

23. Dimeski, G.; Badrick, T.; St John, A. Ion Selective Electrodes (IEs) and Interferences — A Review. *Clin. Chim. Acta* **2010**, *411*, 309–317.

24. Dimeski, G.; et al. Disagreement between Ion Selective Electrode Direct and Indirect Sodium Measurements: Estimation of the Problem in a Tertiary Referral Hospital. *J. Crit. Care* **2012**, *27* (326), e9–16.

25. Chow, E.; Fox, N.; Gama, R. Effect of Low Serum Protein on Na an K Measurement by ISE in Critically Ill Patients. *Br. J. Biomed. Sci.* **2008**, *65*, 128–131.

26. Wennecke, G. *Useful Tips to Avoid Preanalytical Errors in Blood Gas Testing: Metabolites,* January 2004. www.acutecaretesting.org.

27. Stankovic, A. K.; Smith, S. Elevated Serum Potassium Values: The Role of Preanalytical Variables. *Am. J. Clin. Pathol.* **2004**, *121* (Suppl. 1), S105–S112.

28. Don, B. R.; Sebastian, A.; Cheitlin, M.; Christiansen, M.; Schambelan, M. Pseudohyperkalemia Caused by Fist Clenching During Phlebotomy. *N. Engl. J. Med.* **1990**, *322*, 1290–1292.

29. Scott, M.; LeGrys, V. A.; Klutts, S. Electrolytes and Blood Gases. In *Tietz Textbook of Clinical Chemistry and Molecular Diagnostics,* 4th ed.; Elsevier Saunders, 2006.

30. Haverstick, D. M.; Brill, L. B.; Scott, M. G.; Bruns, D. E. Preanalytical Variables in Measurement of Free (Ionized) Calcium in Lithium Heparin-Containing Blood Collection Tubes. *Clin. Chim. Acta* **2009**, *403*, 102–104.

31. Toffaletti, J. G.; Wildermann, R. F. The Effects of Heparin Anticoagulants and Fill Volume in Blood Gas Syringes on Ionized Calcium and Magnesium Measurements. *Clin. Chim. Acta* **2001**, *304*, 147–151.

32. Perovic, A.; Braticevic, M. N. Time-Dependent Variation of Ionized Calcium in Serum Samples. *Biochem. Med.* **2019**, *29* (3), 030708.

33. Desai, T. K.; Carlson, R. W.; Thill-Baharozian, M.; Geheb, M. A. A Direct Relationship between Ionized Calcium and Arterial Pressure Among Patients in an Intensive Care Unit. *Crit. Care Med.* **1988**, *16*, 578–582.

34. Toffaletti, J.; Abrams, B. Effects of In Vivo and In Vitro Production of Lactic Acid on Ionized, Protein-Bound, and Complex-Bound Calcium in Blood. *Clin. Chem.* **1989**, *35*, 935–938.

35. Wang, S.; McDonnell, E. H.; Sedor, F. A.; Toffaletti, J. G. pH Effects on Measurements of Ionized Calcium and Ionized Magnesium in Blood. *Arch. Pathol. Lab Med.* **2002**, *126*, 947–950.

36. Toffaletti, J. G. Use of Novel Preparations of Heparin to Eliminate Interference in Ionized Calcium Measurements: Have All the Problems Been Solved? *Clin. Chem.* **1994**, *40*, 508–509.

37. Ljunghall, S.; Joborn, H.; Benson, L.; Fellstrom, B.; Wide, L.; Akerstrom, G. Effects of Physical Exercise on Serum Calcium and Parathyroid Hormone. *Eur. J. Clin. Invest.* **1984**, *14*, 469–473.

38. Markowitz, M. D.; Arnaud, S.; Rosen, J. F.; et al. Temporal Interrelationships between the Circadian Rhythms of Serum Parathyroid Hormone and Calcium Concentrations. *J. Clin. Endocrinol. Metab.* **1988**, *67*, 1068–1073.

39. Baird, G. S. Ionized Calcium. *Clin. Chim. Acta* **2011**, *412* (9–10), 696–701.

40. Toffaletti, J.; Savory, J.; Gitelman, H. J. Use of Gel Filtration to Examine the Distribution of Calcium Among Serum Proteins. *Clin. Chem.* **1977,** *23,* 2306−2310.
41. Toffaletti, J.; Hammes, M. E.; Gray, R.; Lineberry, B.; Abrams, B. Lactate Measured in Diluted and Undiluted Whole Blood and Plasma: Comparison of Methods and Effect of Hematocrit. *Clin. Chem.* **1992,** *38,* 2430−2434.
42. Astles, R.; Williams, C. P.; Sedor, F. Stability of Plasma Lactate In Vitro in the Presence of Antiglycolytic Agents. *Clin. Chem.* **1994,** *40,* 1327−1330.
43. Calatayud, O.; Tenias, J. M. Effects of Time, Temperature, and Blood Cell Count on Levels of Lactate in Heparinized Whole Blood Gas Samples. *Scand. J. Clin. Lab. Invest.* **2003,** *63,* 311−314.
44. Bazydlo, L. A. L.; Needham, M.; Harris, N. S. Calcium, Magnesium, Phosphate: Review. *Lab. Med.* **2014,** *45* (1). www.labmedicine.com.
45. Ritter, C.; Ghahramani, M.; Marsoner, H. J. More on the Measurement of Ionized Magnesium in Whole Blood. *Scand. J. Clin. Lab. Invest.* **1996,** *56* (Suppl. 224), 275−280.
46. Zoppi, F.; de Gasperi, A.; Guagnellini, E.; et al. Measurement of Ionized Magnesium with AVL 988/4 Electrolyte Analyzer: Preliminary Analytical and Clinical Results. *Scand. J. Clin. Lab. Invest.* **1996,** *56* (Suppl. 224), 259−274.

Chapter 11

Quality control in blood gas and critical care testing

Routine daily quality control on blood gas instruments

Most current blood gas/electrolyte/metabolite analyzers have onboard internal quality control (QC) systems that automatically monitor the stability and performance of reagents, electrodes, and electronics of the analyzer. These control systems make these "hybrid" analyzers well-suited to use in both laboratory and Point-of-Care Testing (POCT) areas. Two such analyzers use easily replaced multiuse packs that contain controls, calibrants, rinse solution, and waste. Some also have the flow system and analyte sensors within the pack. They are described in the following.

One model of a hybrid blood gas analyzer measures pH, pCO_2, pO_2, Na^+, K^+, Cl^-, Ca^{2+}, glucose, lactate, total hemoglobin (Hb), and several forms of Hb including oxy-Hb, carboxy-Hb, and met-Hb. Part of its QC system utilizes four process control solutions (PCS) all contained within a multiuse test pack to monitor the stability of the analytical responses. PCS A and B are run at least every 4 h and after every sample. PCS C is run once per 24 h, and PCS D is run every 12 h. Once the test pack is validated, an active QC program monitors the analytical process before, during, and after sample measurement, with automatic error detection. During the analysis, the sensor response is continually evaluated for abnormal responses that might indicate microclots, microbubbles, or some chemical interferences. If an error is detected in the analysis, the system automatically tries to correct the system error and will either document any corrective actions or disable that test channel if the problem cannot be fixed.

Another brand of hybrid blood gas analyzer measures pH, pCO_2, pO_2, Na^+, K^+, Ca^{2+}, Cl^-, glucose, lactate, bilirubin, and Hb with cooximetry. This analyzer utilizes a replaceable cassette of solid-state amperometric and potentiometric ion-selective sensors, optical pO_2 detection, and spectrophotometric measurements for oximetry. QC solutions are contained within a multiuse pack containing three calibrant solution pouches (with one also used for rinse), one gas mixture pouch, three QC solution pouches, a flow selector, pump tubing, and a waste pouch. To compensate for drift, a calibrant solution is run to bracket every test measurement. Every 8 h, one level of each of three

Blood Gases and Critical Care Testing. https://doi.org/10.1016/B978-0-323-89971-0.00008-2

levels of liquid internal QC solutions is automatically run. All controls enter through the sample probe and flow through the entire system. Clots are usually trapped by a clot catcher near the sample inlet, plus pressure sensors detect any blockage of flow along the analysis path. If abnormal pressure is detected, the analyzer initiates a clot removal process to rid the offending clot from the flow path. To ensure that accurate results are reported, a question mark appears next to any invalid result that may be caused by an inhomogeneous sample, insufficient sample, or some problem with the calibration.

Hand-held portable blood gas analyzers utilize both a liquid calibrant contained within a single-use test card, and an external multiuse electronic simulator card that checks the acceptability of the electronic sensors within the instrument. This simulator is often referred to as providing equivalent quality control (EQC) in place of external control solutions. By itself, it verifies only the electronic function of the analyzer, but does not check the reagent or sensor stabilities of the test card. Every 24 h when cartridges are being tested, the external electronic simulator is used to verify any internal electronic failures. Twice per year, this simulator performs the required thermal probe check. The electronic simulator sends signals that are below and above the measurement ranges of the tests to simulate signals of a test cartridge during an actual analysis of a sample and these simulator results are stored in the instrument's memory. However, these electronic checks typically have almost no variation and do not represent the more realistic precision results obtained with either liquid controls or patient samples.

Another single-use disposable cartridge contains microsensors, a calibrant solution, flow system, and a waste chamber. Sensors for analysis of total CO_2, sodium, potassium, chloride, ionized calcium, glucose, creatinine, urea as blood urea nitrogen (BUN) and hematocrit are available in a variety of panel configurations. Approximately two−three drops (100−150 μL) of anti-coagulated blood are dispensed into the cartridge sample well, which is sealed before inserting the cartridge into the analyzer. Inserting a test card brings the card's sensors into contact with the instruments electrical contacts. This contact activates a heater assembly to bring the measurement region to 37°C. This also opens the fluidic valve in the card to deliver calibrant fluid from the reservoir to the measurement region.

After calibration, and upon a prompt by an LED and an audio beep, the user introduces a blood sample for measurement through the blood sample port to the card's measurement region. When blood contacts the analyte sensors, an electrical signal is generated that is proportional to analyte concentrations in the blood sample.

Individualized Quality Control Plan

Many current blood gas instruments have some type of internal onboard QC. Some analyzers use internal liquid control solutions that flow through the analyzer system making contact with the analyte sensors. Other analyzers use an electronic simulator device that is inserted into the analyzer to check only the electronics of the analyzer. For many years, there has been debate about whether these internal controls fulfill the requirements for external controls as per CLIA 88. For internal liquid controls the issue is whether the daily QC solution is introduced into the system through the sample probe and flows through the entire system, or enters the sample flow path immediately after the sample probe. Manufacturers claim the analysis of these internal controls, along with other quality checks of the system, satisfy the CLIA requirements for external QC.

For hand-held portable analyzers that use an external electronic simulator to verify the electronic functions of the analyzer, a liquid calibrant is contained within the single-use test card. Use of this calibrant provides a type of one-point calibration of the test cards to help ensure acceptable performance of the analyte sensors in the test card.

In 2013, the Centers for Medicare and Medicaid Services issued a "voluntary" option for meeting CLIA quality control standards, which was called Individualized Quality Control Plan (IQCP) that went into effect January 2016 (1). It applied to all labs with analyzers that use EQC. CLIA QC regulations remained the same as published in 2003. All of the preanalytical, analytical, and postanalytical systems requirements in the CLIA regulations remained in effect (2).

Much has been written about IQCP, so this section will only cover brief aspects of IQCP for clinical laboratories. The aim of IQCP was to monitor all phases of testing by developing a plan that assesses potential risks for (1) specimen handling, (2) personnel who test the specimens, (3) instrument reliability, (4) reagent reliability, and (5) environmental factors that could affect results (2). Developing and implementing an IQCP that addresses these potential risks could potentially overcome the limitations of an electronic simulator used as a control device.

The logic of IQCP is apparently that the laboratory runs the risk of error by utilizing internal QC as its external QC requirement. If a blood gas instrument introduces the internal QC sample automatically, it does not check the quality of sample handling by personnel, which is a mandatory area of risk assessment. Because no QC material tests the preanalytic phase of sample transport and handling, the IQCP requires a risk assessment of potential errors during this phase, particularly when time from patient draw to introducing the sample is a critical factor in testing samples, as with blood gas testing. While IQCP attempts to address all phases of testing, it still allows use of electronic simulators as controls, which of themselves provide minimal assurance of test quality.

To be considered an external control material, the control material must have a similar matrix to that of patient specimens, be treated in the same manner as patient specimens, and go through all elements of the analytical process. It must also be a different control or from a different lot number than used to calibrate the instrument (42CFR493.1256(d)(9)). Because electronic simulator controls only verify electronic and mechanical functions of the instrument, they **do not** meet the definition of an external control material.

Laboratories should evaluate their QC processes to determine if they monitor the entire analytical testing process. If the control process does not meet the criteria described for external control materials, the laboratory must either perform additional QC testing, such as analyzing appropriate external control materials once per 8 h, or implement an IQCP to meet requirements for daily QC. If onboard QC solutions meet the above definition of an external control material, an IQCP is not required. The laboratory can periodically verify proper sampling technique through proficiency testing, competency assessment, the use of other types of external control materials, and following processes defined in the manufacturer's instructions.

The minimum guidelines according to CLIA 88 (2003) require laboratories to perform external QC at least one time per 8-h shift *(3)*. In addition to the daily QC requirement for all nonwaived moderate-to-high complexity test systems, the laboratory must also perform and document calibration verification at least twice per year, as well as when any of the following occur: (1) after major maintenance; (2) when replacing significant parts that could affect an instrument's performance; (3) when lot numbers of reagents are changed; and (4) whenever the laboratory identifies a trend or shift in its control material with results falling outside acceptable limits.

With the IQCP regulation now in effect, laboratories should develop a QC plan that includes the use of control materials that resemble patient samples. To eliminate the question of internal QC material qualifying as an external control, some suggest having personnel introduce at least one QC sample every 8-h *(2)*.

Between-lab and between-instrument QC

To verify comparability of multiple instruments testing the same analyte, CAP regulation COM.04250 requires that labs verify comparability of results at least twice a year, preferably using human samples. As part of our routine between-laboratory QC survey, every 2 weeks we pool heparinized blood gas samples destined for discard that remain after testing and reporting *(4)*. Leftover blood samples are randomly selected based on having sufficient volume remaining after testing, with no effort made to select any concentrations of analyte. Before testing, all pooled blood is checked for hemolysis by centrifugation. Twenty-five milliliter of blood is pooled into a plastic medicine cup, mixed, then aliquoted into several 3 mL syringes that are sealed then laid on ice as they are taken on a rotating basis to each laboratory and POCT location for analysis. The results are compiled in a report that is sent to each laboratory for review by the laboratory manager and director.

TABLE 11.1 Acceptable differences from the mean for analytes on blood gas analyzers.

Analyte	Acceptable differences from the mean, with units
pH	± 0.03 units
pCO_2	± 3 mmHg or 7.5%
pO_2	± 5 mmHg or 7.5%
Na	± 3.0 mmol/L
K	± 0.3 mmol/L
Ionized Ca	± 0.04 mmol/L
Glucose	± 6 mg/dL or 10%
Lactate	± 0.4 mmol/L or 10%
Hb	± 0.5 g/dL or 5.0%
%O_2Hb	± 3%

For evaluating results on this between-lab QC program, the mean of all results for each analyte is calculated, with acceptable analytical limits set for each test done on our blood gas analyzers, as shown in Table 11.1. These ranges of acceptable differences for each analyte are based on, but usually more rigorous than, CLSI guidelines *(4,5)*.

Detection of hemolysis in whole blood specimens

As noted in another chapter, detection of hemolysis in anticoagulated blood is a significant concern for blood gas and electrolyte testing, most critically for its effect on potassium results. Companies are apparently trying to develop systems that detect hemolysis for blood gas analyzers. The basic technology may be based on isolating plasma during the measurement process, then using optical methods to detect hemolysis, or using an algorithm based on multiple analyte inputs to predict the degree of hemolysis in whole blood.

References

1. *Individualized Quality Control Plan (IQCP),* December 2019. https://www.cms.gov/ Regulations-and-Guidance/Legislation/CLIA/Individualized_Quality_Control_Plan_IQCP.
2. Jordan, A. *External and Internal QC for Blood Gases.* MLO-Online, July 2016. *h*ttps://www. mlo-online.com/home/article/13008814/external-and-internal-qc-for-blood-gases.
3. American College of Physicians. *CLIA and Your Laboratory: A Guide for Physicians and Their Staff,* November 2014. https://www.acponline.org/system/files/documents/running_ practice/mle/clia-and-your-lab.pdf.

4. Toffaletti, J. G.; Buckner, K. A.; Liu, B.; et al. Retrospective analysis of quality control data using pooled blood to compare agreement within different brands of blood gas analyzers. *J. Appl. Lab. Med.* **2021,** jfab029. https://doi.org/10.1093/jalm/jfab029.

5. Blood Gas and pH Analysis and Related Measurements, Approved Guideline. In *CLSI Document C46-A2,* 2nd ed.; Clinical and Laboratory Standards Institute: Wayne, PA, February 2009.

Chapter 12

Models for point-of-care testing of critical care analytes

Introduction

Point-of-care testing (POCT) is now common in many near-patient and critical care settings. For blood gas and electrolyte testing, this includes operating rooms (ORs), intensive care units (ICUs), cardiac catheterization labs (CCL), emergency departments (ED), and many primary care clinic settings. POCT has several distinct benefits, such as minimizing or eliminating specimen transportation and processing which have the added benefits of minimizing preanalytical effects and providing faster turnaround times (TAT). Some systems offer a variety of test cartridges that provide a flexible test menu. More rapid test results potentially allow more prompt medical decisions, which can lead to improved patient outcomes, operational efficiencies, and patient satisfaction. For a variety of reasons, POCT devices usually require less blood volume especially compared to central laboratory requirements and can be especially attractive in pediatric areas (1). As POCT is not usually performed by trained laboratory personnel, maintaining regulatory compliance and quality assurance with POCT are challenges that require continual surveillance.

Most current POCT devices can be interfaced to both laboratory information systems (LIS) and electronic medical record systems (EMR). For today's current high test volumes and the need to minimize transcription errors, connection to information systems is essential. Well-functioning information systems can automatically transfer (download) results from the analyzer and display them to the physician. This ensures both accuracy and rapid delivery of results to multiple caregivers.

An issue that sometimes arises is the potential for POCT to replace central laboratory (CL) testing. POCT has, and will, replace some CL testing, with POCT accounting for 10%–20% of clinical laboratory testing, depending on the location. This is true where rapid results are essential for urgent decisions, and helpful in nonurgent settings such as clinics that potentially allow the physician to evaluate the results and discuss with the patient while still present.

Blood Gases and Critical Care Testing. https://doi.org/10.1016/B978-0-323-89971-0.00003-3

In addition, POCT can replace near-patient laboratories that are extremely inefficient. However, the sheer test volume handled by most central laboratories cannot be replaced by POCT, at least with present technology. Most, if not all, POCT devices are not equipped to handle large test volumes and generally would be far more expensive to operate.

Importance of POCT in answering clinical needs

In emergency and critical care settings, such as ICUs, ORs, EDs, and CCLs, blood gas and electrolytes can change rapidly and often require equally rapid clinical decisions. In these settings, a variety of patients may be seen, including burn, trauma, chest pain, sepsis, and stroke patients, and those on ventilators *(2)*. Because these situations usually require continuous monitoring, POCT is well suited to provide rapid results that aid in urgent clinical decisions that may lead to improved patient outcomes *(2−4)*. A recent document notes that when the laboratory TAT exceeds 25% of the desired decision time, POCT should be considered *(4)*.

Blood gases. Because pH, pCO_2, and pO_2 can change suddenly with potentially life-threatening consequences, blood gas results are virtually always required immediately. Blood gas results by POCT can be most helpful in ICUs and EDs for patients under anesthesia or on ventilation, or those with cardiac failure, hypoxemia, acidosis, sepsis, or any shock state *(3)*.

Sodium, potassium, chloride. Detecting, monitoring, and treating abnormal electrolyte concentrations are critically important in many patients. As noted in a review *(3)*, many studies conclude that rapid results provided by POCT reduce therapeutic intervention time compared to TAT for results from central laboratories.

Ionized calcium. In ICU settings, prompt ionized calcium results are especially helpful for patients with sepsis, hypocalcemia, arrhythmias related to hyperkalemia, hypotension, heart failure, shock, and burns *(5)*. Ionized calcium concentrations below 0.70 mmol/L have been associated with higher morbidity and mortality *(5)*.

Lactate. Many studies have shown that elevated blood lactate concentrations are associated with increased mortality and that rapid results by POCT can be beneficial *(3,4)*. Critical concentrations of lactate differ, with some finding that concentrations >4.0 mmol/L were associated with a much higher risk of mortality in sepsis *(6)*, and others finding that lactates >2.0 mmol/L on admission to the ICU was associated with higher mortality *(7)*.

Types of POCT analyzers

Stationary "hybrid" analyzers used in POCT areas. Hybrid analyzers are those with qualities that are suitable for use in both clinical laboratories and in certain POCT locations. Typically, maintenance involves simply replacing one or only a few test packs that contain reagents, controls, calibrants, sampling and flow systems, and sometimes electrodes and optical detection systems in packs that are relatively easily replaced. Current hybrid analyzers typically give results in 1 min or less and sample volume requirements are similar to and sometimes less than for portable analyzers. They have both ion-selective-electrode (ISE) and optical detection systems to measure the full panel of "blood gas" tests, including electrolytes, glucose, lactate, and hemoglobin with variants. Because these analyzers cost more and are typically more expensive to operate for lower test volumes, they are more suited for areas with relatively higher test volumes.

Portable analyzers used for POCT. Portable analyzers are handheld and are easily transported or carried for POC testing. They may be either located in a docking station for use when such testing is needed or carried continually by those who do more frequent blood gas testing, such as respiratory therapists. These instruments have the distinct advantages of portability, small sample volume requirement, flexible test menu (although sometimes requiring multiple test cartridges), and usually lower cost per instrument. For small test volumes, they can be less costly per test than hybrid systems, but the cost per reportable result is usually higher for large test volumes. Because most use single-use test cartridges, the overall time to obtain the final result is typically longer than with a hybrid analyzer, so portable devices are not well-suited for handling large test volumes. Analytical quality is quite good, but usually a notch below the standards of a bench or hybrid analyzer. Most or all of the handheld portable systems are not able to directly measure the cooximetry tests (Hb, O_2Hb, CO-Hb, and met-Hb). Instead, hematocrit is measured by impedance that is used to calculate Hb, For some complete test panels, two cartridges may be needed.

Selecting an analyzer for POCT

Many factors are involved in selecting a POC analyzer that fulfills the clinical and operational needs. These include the following *(4)*:

- Analyzer quality
- Staffing needs
- Time to the result and time to recycle to the next sample
- Material and storage requirements
- Flexibility of information systems to meet testing and monitoring needs
- Cost justification
- Complexity and ease of use of the test system by nonlaboratory personnel.
- Ease of testing QC material
- Size of the analyzer

Analyzer quality for POCT

The reliability and performance characteristics of a POC analyzer must suit the clinical needs of the physician in monitoring and treating the patient. While POC devices provide quality results, they often use very different technology than core lab instruments and generally have some analytical differences (biases, imprecision, interferences) relative to lab instruments.

Quality begins with the manufacturer of the devices and must meet standards of safety, analytical quality, reliability, and relative ease of use among different testing personnel. Operator performance is also highly important so that a good device used with poor technique will likely give an unreliable result. The CDC studied the use of glucometers in various settings and found the following *(8)*:

- Best performance was POCT done by medical technologists associated with hospital labs
- Worst POCT performance was by nonlab personnel in physician offices;
- Intermediate performance was from nonlab personnel trained by laboratory or laboratory-trained users.

To summarize, the laboratory should be responsible for oversight of POCT to ensure quality and to educate users to understand that quality includes the following:

- Appropriate planning
- Implementing and selecting appropriate POCT systems that fulfill the clinical needs
- Ensuring consistent analysis of quality control material and timely evaluation of analyzer functions
- Timely and successful results on external quality assessment by proficiency testing
- Training and continual education of POCT operators
- Rapid and easily documented clinical results
- Selecting analyzers that have the same or similar reference intervals (ranges) as those for the laboratory

Information connectivity and data management

Information technology related to POCT is the science of how information and data are acquired, processed and organized, stored, and transmitted or communicated for both immediate and retrospective clinical interpretation. Data management systems automate the ordering and billing of results, monitor QC data for trends or shifts, report patient results, detect and notify laboratory staff of error codes, and document competency training of operators *(4)*.

Indeed, functional interfacing of POCT devices to LIS and the EMR are essential in hospital settings. In evaluating a POC analyzer, the laboratory staff must determine and understand the ability of the instrument system to interface with the existing LIS and EMR, or what middleware system must be purchased for such compatibility. Automatic transfer of results virtually eliminates transcription errors, which would be an overwhelming problem in a POC setting with end users entering results. With either portable or stationary "hybrid" analyzers, wireless connection is highly desirable, or at least a docking station that will automatically download any results stored in the analyzer. If wireless connectivity is not available, stationary analyzers will require "hard-wiring" to information systems.

Another developing area is transmitting results directly to the patient as a consumer. This is called direct-to-consumer (DTC) laboratory testing that permits consumers to order laboratory tests directly from a laboratory without necessarily having to work with a healthcare provider. Test results given to the consumer may be used to monitor an existing health condition, identify a previously unknown medical disorder, or provide personal health data. While DTC laboratory testing aims to increase individuals' engagement in managing their healthcare, the critical nature of blood gases, the specimen collection and handling requirements, and the moderate complexity of blood gas and electrolyte testing are not conducive to DTC testing. There are many issues with DTC testing that are discussed in a position statement *(9)*.

Cost analysis for handheld versus hybrid analyzers

This section will analyze the costs for hypothetical handheld (HH) analyzers versus hypothetical hybrid (Hyb) analyzers. Several assumptions will be made, as this cost analysis is an example only, so other costs and numbers can be substituted for other situations. This is the model for purchasing the analyzers outright, then paying for cartridges or reagents as needed. Certainly, analyzers and reagents are often grouped together in a sales contract, so that would require a different cost analysis. Importantly, this analysis assumes only one cartridge would be used for the HH analyzer test menu. The test menus are typically different between these types of analyzers, so if more than blood gases are needed, two cartridges may be required for the test panel on the HH analyzers. On the other hand, the cost per test for the Hyb analyzers may be higher for low test volumes versus high test volumes. Also, this assumes 100% efficiency for each system, with no wastage because of any repeat testing required, plus no QC or calibration tests are included.

This analysis shown in Table 12.1 and Fig. 12.1 indicates that the "break even" test volumes are about 2700/year if one analyzer each is used, and about 1500/year if four HH analyzers and two Hyb analyzers are used. As noted, there are many other factors to consider, including test menus and number of cartridges needed for a desired panel, ease of use, sample volume, supplies for one or multiple test platforms, etc.

TABLE 12.1 Yearly costs for handheld (HH) and hybrid (Hyb) POCT systems for yearly test volumes.

	Costs for one HH analyzer	Cost for four HH analyzers	Costs for one Hyb analyzer	Cost for two Hyb analyzers
Analyzer cost	$8000	$32,000	$19,000	38,000
Cost per test panel	$6.00	$6.00	$2.00	$2.00
500 tests/ year	11,000	35,000	20,000	39,000
1000 tests/year	14,000	38,000	21,000	40,000
2000 tests/year	20,000	44,000	23,000	42,000
5000 tests/year	38,000	62,000	29,000	48,000
10,000 tests/year	68,000	92,000	39,000	58,000
20,000 tests/year	128,000	152,000	59,000	78,000

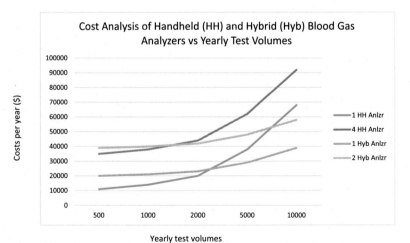

FIGURE 12.1 Plots of yearly costs for hand-held (HH) and hybrid (Hyb) POCT systems for yearly test volumes. The "break even" test volumes are approximately 2700 tests/year for one Hand Held Analyzer (HH Anlzr) vs one Hybrid Analyzer (Hyb Anlzr), and approximately 1500 tests/year for four HH Anlzrs vs two Hyb Anlzrs. Above those test volumes the Hand Held Analyzers would cost more per test. *This is an original figure.*

Quality control techniques for POC testing

Quality control is covered in another chapter, but briefly, blood gas testing is classified as moderately complex, so Clinical Laboratory Improvement Amendments (CLIA) section 493.1267 states that blood gas testing requires analyzing at least one sample of QC material every 8 h of testing and three levels (low, normal, and high) of QC every 24 h of testing *(10)*. Since 2016, if QC testing frequency or control materials do not meet these standards, an IQCP must be developed *(11)*. For waived testing, the lab need only follow the manufacturer's instructions.

Meeting compliance requirements by regulatory agencies

Major parts of POCT regulatory requirements focus on training and competency of testing personnel and verifying adherence to the manufacturer's instructions for each test. The latter is especially important because waived or moderately complex laboratory methods become highly complex tests if the testing deviates from the FDA-approved manufacturer's protocol. Because high complexity laboratory tests must be performed by personnel meeting higher CLIA qualifications, such testing is usually not possible at POC.

Competency assessment of testing personnel led the list of CAP deficiencies for 2016, followed by instrument correlations/method comparisons, review of procedure manuals by the laboratory staff/medical director, evaluation of proficiency testing results, and review of equipment function checks/maintenance records *(12)*. Proper reagent labeling, updating the activity menu, and documentation of method validation also resulted in a significant number of deficiencies (See Table 12.2).

Training and continuing compliance

The initial training of the end users is challenging but essential, not only for quality results, but to maintain proper instrument function and cleanliness, documentation, and compliance with regulations. The laboratory is responsible for stressing the importance of quality, maintaining quality, and ensuring that the ongoing compliance requirements and responsibilities are adhered to by the end users.

Maintaining continual analytical quality and compliance training for the users is essential in any POCT program. There are many models of training that can work well, depending on the number of operators to be trained, the number of designated laboratory trainers, and the number of trained "superusers" among the clinical staff. Maintaining documentation of approved users requires dates of initial and continual training, which are typically after the first 6 months, then at every 12-month interval following the initial

TABLE 12.2 Most frequently cited deficiencies found at CAP inspections.

Rank of most frequently cited deficiencies by CAP	Type of deficiency
1	Testing personnel competency
2	Providing an accurate test menu
3	Having a procedure manual
4	Correlations between instruments
5	Procedures reviewed by lab director
6	Complete maintenance records
7	Labeling of reagents and expired reagents
8	Evaluation and signature review of proficiency testing
9	Checking test accuracy 2× per year
10	Approval signature of method validation

training. Documentation of acceptable education is now required, which can be very challenging for a variety of reasons, including that perfusionists, nurses, and other POCT personnel often feel that, because they have already passed the education requirements for their job, their credentials are being questioned. While this process can be challenging when initially getting education documents from current users, getting such documents from new users soon after they are hired is usually an easier process.

Case example: decisions in implementing a POCT system

Before implementing a POC testing system, the overall benefits must be weighed versus the costs and efforts both to implement and to maintain the system and process. A guiding principle of POCT is that each situation will require a specific solution. Therefore, this section is intended to provide general guidance in a case study approach to implementing POC blood gas and electrolyte testing.

What needs are not met by central lab testing?

For many years, a full-service high-volume pediatric laboratory located amid various pediatric ICUs was able to serve the needs of several ICUs: neonatal (NICU), pediatric (PICU), pediatric cardiac (PCICU), and labor and delivery. Both the PCICU and PICU moved to a new bed tower that was more distant from the Pediatric Laboratory. Because the PCICU accounted for a high volume of blood gas testing, two new hybrid blood gas analyzers were purchased and located at stations in the PCICU.

What will be improved by implementing POCT?

Having two blood gas analyzers to provide rapid results in the PCICU area should provide results similar to or even faster than in the former location close to the full-service laboratory. By using the same brand of hybrid analyzers as those in the clinical pediatric laboratory, comparability of results between the laboratory and POC analyzers was assured.

What are the training needs and additional staffing needs of both the POC users and the laboratory?

This was a significant issue and required a very large time commitment by the laboratory staff for training. Because POCT needed to be done 24/7, it required a very large involvement of the nursing staff. In our POCT model for the PCICU, we initially estimated that 25 users would be required to ensure 24/7 testing in the unit. For this, we enlisted one very experienced laboratory trainer with the ability and interactive skills for this job. Initially, this person trained 15 respiratory therapists (RT), but as this evolved it ended up with 10 RTs and 34 nurses, one nurse practitioner and one MD who was highly committed to POCT in this area. Clearly, this process evolved as the needs became more apparent, which highlights the importance in POCT of continually evaluating if processes fulfill needs, and adjusting processes as necessary.

End-user acceptability of responsibilities. Is there buy-in not only from the managers of the care units, but from those who will do the testing?

Although the physician director and nurse managers of the unit were clearly in favor of POCT, the enthusiasm of the nursing staff was more variable. This required considerable education and patience with the staff. However, as training progressed, the nurses became more comfortable with and accepting of the analyzers and POCT.

Continual communication with users and having a supportive point person

The physician director of the PCICU was highly supportive of POCT for blood gas and electrolytes. That provided the momentum for evaluating the costs and labor needed to provide POCT. This involved several preliminary meetings with laboratory and clinical staff to discuss the needs, concerns, and challenges to address in developing plans to make this POCT a success.

What are the start-up and operating costs? How complex and technically difficult is the device for POC testing? Does the system store names of approved operators and require an operator ID to prevent nonauthorized operators?

While the most important considerations were for the analyzers to provide quality results with ease of use, require small sample volumes, and ideally require minimum additional supplies to add to the laboratory inventory, costs were also important to evaluate. For this PCICU the blood gas test volume was approximately 20,000 per year. With this high test volume, and based on prior familiarity with the analyzer, two stationary hybrid blood gas analyzers of the same type as used in the pediatric laboratory were selected for the PCICU. The analyzers were easy to use, required a very small blood volume, and provided results in less than 1 min. They also prevent unauthorized use by requiring trained and approved operators to enter their unique ID to use the analyzer.

What are the space needs, including electrical and informatics connections?

These analyzers have a small footprint, which would require relatively little space. However, analyzers require space for computer terminals, a nearby phone, a nearby sink, storage (possibly refrigerated) of reagents and supplies such as gloves, and a biohazard waste container. Being a new building, the electrical, phone, and computer/information hookups required only minor renovations. These analyzers also had to be in a location convenient for testing, which can be challenging as such space may be desirable for other functions. In our case, blood gases were a high priority supported by the physician director and nurse manager of the unit, so suitable locations were found.

Who will do maintenance and troubleshooting?

We felt the most reliable solution would be for the laboratory staff to perform all maintenance and certainly any troubleshooting. This also included maintaining the inventory of supplies. As troubleshooting was clearly the responsibility of the laboratory, it was essential for the laboratory to have the ability to monitor the performance and onboard supplies of the analyzers. As with most current analyzers, these have interfaced middleware programs that can remotely monitor analyzer performance and onboard supplies for the instrument. This real-time monitoring of our POC blood gas analyzers allowed the laboratory to quickly become aware of any deterioration of performance and shortage of supplies.

This POCT program is a "work in progress" as are most POCT programs. Overall, this process has been a success with approximately 90% of blood gas tests now done by the PCICU staff.

References

1. Patel, K.; Suh-Lailam, B. B. Implementation of Point-of-Care Testing in a Pediatric Healthcare Setting. *Crit. Rev. Clin. Lab. Sci.* **2019,** *56* (4), 239−246.
2. D'Orazio, P.; Fogh-Andersen, N.; Okorodudu, A.; Shipp, G.; Shirey, T.; Toffaletti, J. Chapter 5: Critical Care. In *Evidence Based Practice for Point-of-Care Testing: A National Academy of Clinical Biochemistry Laboratory Medicine Practice Guideline;* Nichols, J. H., Ed.; AACC Press: Washington DC, 2006; pp 30−43.
3. Kapoor, D.; Srivastava, M.; Singh, P. Point of Care Blood Gases with Electrolytes and Lactates in Adult Emergencies. *Int. J. Crit. Illn. Inj. Sci.* **2014,** *4* (3), 216−222.
4. Nichols, J. H.; Alter, D.; Chen, Y.; Isbell, T. S.; Jacobs, E.; Moore, N.; et al. AACC Guidance Document on Management of Point-of-Care Testing. [Epub]. *J. Appl. Lab. Med.* **2020** **2020**, (4). https://doi.org/10.1093/jalm/jfaa059.
5. Zivan, J. R.; Gooley, T.; Zager, R. A.; Ryan, M. J. Hypocalcemia: A Pervasive Metabolic Abnormality in the Critically Ill. *Am. J. Kidney Dis.* **2001,** *37,* 689−698.
6. Soliman, H. M.; Vincent, J.-L. Prognostic Value of Admission Serum Lactate Concentration in Intensive Care Unit Patients. *Acta Clin. Belg.* **2010,** *65,* 176−181.
7. Moore, C. C.; Jacob, S. T.; Pinkerton, R.; Meya, D. B.; Mayanja-Kizza, H.; Reynolds, S. J. Point-of-Care Lactate Testing Predicts Mortality of Severe Sepsis in a Predominantly HIV type 1-Infected Patient Population in Uganda. *Clin. Infect. Dis.* **2008,** *46,* 215−222.
8. Kohn, L. T., Corrigan, J. M., Donaldson, M. S., Eds. *To Err is Human: Building a Safer Health System. Committee on Quality of Health Care in America. Institute of Medicine;* National Academy Press: Washington, DC, 2000.
9. AACC Position Statement; *Direct-to-Consumer Laboratory Testing,* December 2019. https://www.aacc.org/advocacy-and-outreach/position-statements/2019/direct-to-consumer-laboratory-testing#.
10. Ehrmeyer, S. S. *The Importance of Quality Control (QC) to Quality Blood Gas Testing,* September 2012. www.acutecaretesting.org.
11. *Individualized Quality Control Plan (IQCP),* December 2019. https://www.cms.gov/Regulations-and-Guidance/Legislation/CLIA/Individualized_Quality_Control_Plan_IQCP.
12. Chittiprol, S.; Bornhorst, J.; Kiechle, F. L. *Top Laboratory Deficiencies across Accreditation Agencies;* Clinical Laboratory News, July 2018. https://www.aacc.org/cln/articles/2018/july/top-laboratory-deficiencies-across-accreditation-agencies.

Index

Printed in the United States
by Baker & Taylor Publisher Services